Food Engineering

Food Engineering

Edited by
Keegan Mason

Larsen & Keller
www.larsen-keller.com

Food Engineering
Edited by Keegan Mason
ISBN: 978-1-63549-122-7 (Hardback)

Larsen & Keller
Published by Larsen and Keller Education,
5 Penn Plaza,
19th Floor,
New York, NY 10001, USA

Cataloging-in-Publication Data

Food engineering / edited by Keegan Mason.
p. cm.
Includes bibliographical references and index.
ISBN 978-1-63549-122-7
1. Food. 2. Food--Technological innovations. 3. Food industry and trade. I. Mason, Keegan.
TP370 .F66 2017
664--dc23

Printed and bound in China.

For more information regarding Larsen and Keller Education and its products, please visit the publisher's website www.larsen-keller.com

Table of Contents

Preface

This book outlines the processes and applications of food engineering in detail. It talks extensively about the different branches that fall under this category. Food engineering, as a branch of physical sciences, is used in food industry for food processing, building food machinery, ingredient manufacturing and instrumentation, etc. Food technology is an integral branch that ensures food safety and regulation through the application of technologies that standardize food manufacturing process. The topics included in this book on food engineering are of utmost significance and are bound to provide incredible insights to the readers. The text will serve as a valuable source of reference for those interested in this field and related branches.

Given below is the chapter wise description of the book:

Chapter 1- Most food available in the market are processed before it becomes available to the consumer. Food engineering a specialized sub-field of agricultural engineering and it studies the application of the principles of engineering to the production and distribution of food products. This introductory chapter ensures that the reader is provided with comprehensive background information about food engineering.

Chapter 2- The food processing industry is large and is primarily concerned with converting raw food into marketable food products and hygienically packaging them. Food science as a discipline uses the principles of engineering, physical sciences and biology to create newer and better food processing methods, understand the spoilage of food and to bring improvements in the food packaged. This chapter explores the subjects of food science and food technology and its future.

Chapter 3- Food is grown, transported for processing, processed and finally packaged after which it is marketed to the consumer. An aspect that is crucial for the food industry is safe handling and storage of food. Food science and engineering comprises multiple aspects such as food processing, food preservation, food packaging, food marketing and storage. This chapter studies these aspects, the crucial roles they play and the techniques used in each.

Chapter 4- Food chemistry studies the chemical reactions that take place in food and analyses the interaction of the various components of food. It also studies any change that takes place in food composition like the coagulation of milk into cheese or the browning of cut apples. This section provides valuable information on food coloring, food additives or preservatives, flavors and the nutrient content of food. The chapter elucidates the crucial theories and principles of food chemistry.

Chapter 5- Adulterants in food are unwanted substances that reduce the quality of food or make it harmful. They contaminate the food and make it unfit for consumption. Some adulterants are added on purpose to act as fillers in expensive food products. This chapter discusses the various types of adulterants such as aldicarb, lead and melamine.

Chapter 6- Genetically modified food is produced by genetically modified organisms (GMOs) to allow domination of certain genetic traits and cause recession of unwanted ones. GMOs have ensured higher nutrient value, better resistance to dry conditions, frost, pests, insects etc. as well as delayed or advanced ripening among other food products. Though GMOs have a multitude of benefits and enable scientists to produce crops that have higher yield potential, their use is still controversial as they can wipe out indigenous species of plants and their future safety for the environment and humans is still questionable.

Chapter 7- Food fortification is the addition of nutrients like minerals and vitamins into processed food to help tide over any nutritional deficiencies that could exist or to make food more commercially viable. This chapter contains information on the fortification of food and the common types of fortification. It also has a section on dietary supplements and nutraceuticals. The topics discussed in the chapter are of great importance to broaden the existing knowledge on food engineering.

At the end, I would like to thank all those who dedicated their time and efforts for the successful completion of this book. I also wish to convey my gratitude towards my friends and family who supported me at every step.

Editor

1

Introduction to Food Engineering

Most food available in the market are processed before it becomes available to the consumer. Food engineering a specialized sub-field of agricultural engineering and it studies the application of the principles of engineering to the production and distribution of food products. This introductory chapter ensures that the reader is provided with comprehensive background information about food engineering.

Food Engineering

Bread factory in Germany

Food engineering is a multidisciplinary field of applied physical sciences which combines science, microbiology, and engineering education for food and related industries. Food engineering includes, but is not limited to, the application of agricultural engineering, mechanical engineering and chemical engineering principles to food materials. Food engineers provide the technological knowledge transfer essential to the cost-effective production and commercialization of food products and services. Physics, chemistry, and mathematics are fundamental to understanding and engineering products and operations in the food industry.

Food engineering encompasses a wide range of activities. Food engineers are employed in food processing, food machinery, packaging, ingredient manufacturing, instrumentation, and control. Firms that design and build food processing plants, consulting firms, government agencies, pharmaceutical companies, and health-care firms also employ food engineers. Specific food engineering activities include:

- Drug/Food Products;
- design and installation of food/biological/pharmaceutical production processes;
- design and operation of environmentally responsible waste treatment systems;
- marketing and technical support for manufacturing plants.

Topics in Food Engineering

In the development of food engineering, one of the many challenges is to employ modern tools, technology, and knowledge, such as computational materials science and nanotechnology, to develop new products and processes. Simultaneously, improving quality, safety, and security remain critical issues in food engineering study. New packaging materials and techniques are being developed to provide more protection to foods, and novel preservation technology is emerging. Additionally, process control and automation regularly appear among the top priorities identified in food engineering. Advanced monitoring and control systems are developed to facilitate automation and flexible food manufacturing. Furthermore, energy saving and minimization of environmental problems continue to be important food engineering issues, and significant progress is being made in waste management, efficient utilization of energy, and reduction of effluents and emissions in food production.

Typical Topics Include:

- Advances in classical unit operations in engineering applied to food manufacturing
- Progresses in the transport and storage of liquid and solid foods
- Developments in heating, chilling and freezing of foods
- Advanced mass transfer in foods
- New chemical and biochemical aspects of food engineering and the use of kinetic analysis
- New techniques in dehydration, thermal processing, non-thermal processing, extrusion, liquid food concentration, membrane processes and applications of membranes in food processing
- Shelf-life, electronic indicators in inventory management, and sustainable technologies in food processing
- Modern packaging, cleaning, and sanitation technologies.
- Development of sensors systems for quality and safety assessment

Food

Various foods

Foods from plant sources

Food is any substance consumed to provide nutritional support for the body. It is usually of plant or animal origin, and contains essential nutrients, such as fats, proteins, vitamins, or minerals. The substance is ingested by an organism and assimilated by the organism's cells to provide energy, maintain life, or stimulate growth.

Historically, people secured food through two methods: hunting and gathering and agriculture. Today, the majority of the food energy required by the ever increasing population of the world is supplied by the food industry.

Food safety and food security are monitored by agencies like the International Association for Food Protection, World Resources Institute, World Food Programme, Food and Agriculture Organization, and International Food Information Council. They address issues such as sustainability, biological diversity, climate change, nutritional economics, population growth, water supply, and access to food.

The right to food is a human right derived from the International Covenant on Economic, Social and Cultural Rights (ICESCR), recognizing the "right to an adequate standard of living, including adequate food", as well as the "fundamental right to be free from hunger".

Food Sources

Most food has its origin in plants. Some food is obtained directly from plants; but even animals that are used as food sources are raised by feeding them food derived from plants. Cereal grain is a staple food that provides more food energy worldwide than any other type of crop. Corn (maize), wheat, and rice – in all of their varieties – account for 87% of all grain production worldwide. Most of the grain that is produced worldwide is fed to livestock.

Some foods not from animal or plant sources include various edible fungi, especially mushrooms. Fungi and ambient bacteria are used in the preparation of fermented and pickled foods like leavened bread, alcoholic drinks, cheese, pickles, kombucha, and yogurt. Another example is blue-green algae such as Spirulina. Inorganic substances such as salt, baking soda and cream of tartar are used to preserve or chemically alter an ingredient.

Plants

Many plants and plant parts are eaten as food and around 2,000 plant species are cultivated for food. Many of these plant species have several distinct cultivars.

Seeds of plants are a good source of food for animals, including humans, because they contain the nutrients necessary for the plant's initial growth, including many healthful fats, such as Omega fats. In fact, the majority of food consumed by human beings are seed-based foods. Edible seeds include cereals (corn, wheat, rice, et cetera), legumes (beans, peas, lentils, et cetera), and nuts. Oilseeds are often pressed to produce rich oils - sunflower, flaxseed, rapeseed (including canola oil), sesame, et cetera.

Seeds are typically high in unsaturated fats and, in moderation, are considered a health food, although not all seeds are edible. Large seeds, such as those from a lemon, pose a choking hazard, while seeds from cherries and apples contain cyanide which could be poisonous only if consumed in large volumes.

Fruits are the ripened ovaries of plants, including the seeds within. Many plants and animals have coevolved such that the fruits of the former are an attractive food source to the latter, because animals that eat the fruits may excrete the seeds some distance away. Fruits, therefore, make up a significant part of the diets of most cultures. Some botanical fruits, such as tomatoes, pumpkins, and eggplants, are eaten as vegetables.

Vegetables are a second type of plant matter that is commonly eaten as food. These include root vegetables (potatoes and carrots), bulbs (onion family), leaf vegetables (spinach and lettuce), stem vegetables (bamboo shoots and asparagus), and inflorescence vegetables (globe artichokes and broccoli and other vegetables such as cabbage or cauliflower).

Animals

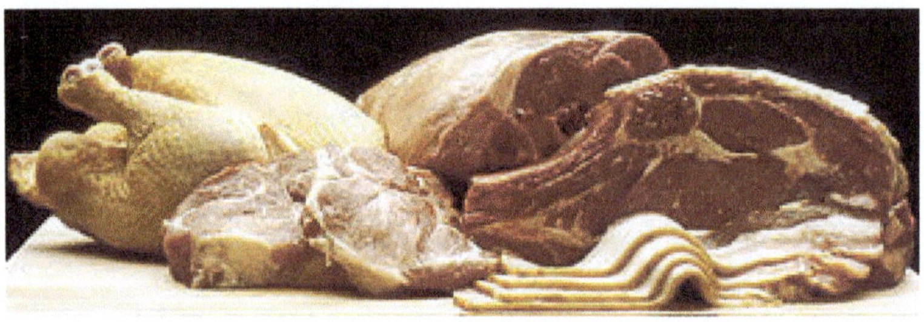

Various raw meats

Animals are used as food either directly or indirectly by the products they produce. Meat is an example of a direct product taken from an animal, which comes from muscle systems or from organs.

Food products produced by animals include milk produced by mammary glands, which in many cultures is drunk or processed into dairy products (cheese, butter, etc.). In addition, birds and other animals lay eggs, which are often eaten, and bees produce honey, a reduced nectar from flowers, which is a popular sweetener in many cultures. Some cultures consume blood, sometimes in the form of blood sausage, as a thickener for sauces, or in a cured, salted form for times of food scarcity, and others use blood in stews such as jugged hare.

Some cultures and people do not consume meat or animal food products for cultural, dietary, health, ethical, or ideological reasons. Vegetarians choose to forgo food from animal sources to varying degrees. Vegans do not consume any foods that are or contain ingredients from an animal source.

Production

A tractor and chaser bin

Most food has always been obtained through agriculture. With increasing concern over both the methods and products of modern industrial agriculture, there has been a growing trend toward sustainable agricultural practices. This approach, partly fueled by consumer demand, encourages biodiversity, local self-reliance and organic farming methods. Major influences on food production include international organizations (e.g. the World Trade Organization and Common Agricultural Policy), national government policy (or law), and war.

In popular culture, the mass production of food, specifically meats such as chicken and beef, has come under fire from various documentaries, most recently Food, Inc, documenting the mass slaughter and poor treatment of animals, often for easier revenues from large corporations. Along with a current trend towards environmentalism, people in Western culture have had an increasing trend towards the use of herbal supplements, foods for a specific group of people (such as dieters, women, or athletes), functional foods (fortified foods, such as omega-3 eggs), and a more ethnically diverse diet.

Several organisations have begun calling for a new kind of agriculture in which agroecosystems provide food but also support vital ecosystem services so that soil fertility and biodiversity are maintained rather than compromised. According to the International Water Management Institute and UNEP, well-managed agroecosystems not only provide food, fiber and animal products, they also provide services such as flood mitigation, groundwater recharge, erosion control and habitats for plants, birds fish and other animals.

Taste Perception

Animals, specifically humans, have five different types of tastes: sweet, sour, salty, bitter, and umami. As animals have evolved, the tastes that provide the most energy (sugar and fats) are the most pleasant to eat while others, such as bitter, are not enjoyable. Water, while important for survival, has no taste. Fats, on the other hand, especially saturated fats, are thicker and rich and are thus considered more enjoyable to eat.

Sweet

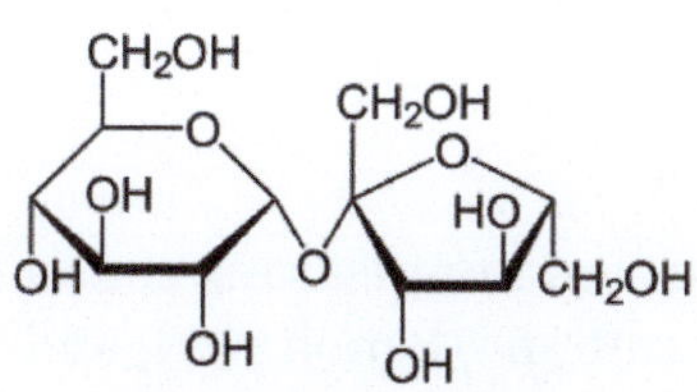

Structure of sucrose

Generally regarded as the most pleasant taste, sweetness is almost always caused by a type of simple sugar such as glucose or fructose, or disaccharides such as sucrose, a molecule combining glucose and fructose. Complex carbohydrates are long chains and thus do not have the sweet taste. Artificial sweeteners such as sucralose are used to mimic the sugar molecule, creating the sensation of sweet, without the calories. Other types of sugar include raw sugar, which is known for its amber color, as it is unprocessed. As sugar is vital for energy and survival, the taste of sugar is pleasant.

The stevia plant contains a compound known as steviol which, when extracted, has 300 times the sweetness of sugar while having minimal impact on blood sugar.

Sour

Sourness is caused by the taste of acids, such as vinegar in alcoholic beverages. Sour foods include citrus, specifically lemons, limes, and to a lesser degree oranges. Sour is evolutionarily significant as it is a sign for a food that may have gone rancid due to bacteria. Many foods, however, are slightly acidic, and help stimulate the taste buds and enhance flavor.

Salty

Salt mounds in Bolivia

Saltiness is the taste of alkali metal ions such as sodium and potassium. It is found in almost every food in low to moderate proportions to enhance flavor, although to eat pure salt is regarded as highly unpleasant. There are many different types of salt, with each having a different degree of saltiness, including sea salt, fleur de sel, kosher salt, mined salt, and grey salt. Other than enhancing flavor, its significance is that the body needs and maintains a delicate electrolyte balance, which is the kidney's function. Salt may be iodized, meaning iodine has been added to it, a necessary nutrient that promotes thyroid function. Some canned foods, notably soups or packaged broths, tend to be high in salt as a means of preserving the food longer. Historically salt has long been used as a meat preservative as salt promotes water excretion. Similarly, dried foods also promote food safety.

Bitter

Bitterness is a sensation often considered unpleasant characterized by having a sharp, pungent taste. Unsweetened dark chocolate, caffeine, lemon rind, and some types of fruit are known to be bitter.

Umami

Umami, the Japanese word for delicious, is the least known in Western popular culture but has a long tradition in Asian cuisine. Umami is the taste of glutamates, especially monosodium glutamate (MSG). It is characterized as savory, meaty, and rich in flavor. Salmon and mushrooms are foods high in umami.

Cuisine

Typical Assyrian cuisine

Many cultures have a recognizable cuisine, a specific set of cooking traditions using various spices or a combination of flavors unique to that culture, which evolves over time. Other differences include preferences (hot or cold, spicy, etc.) and practices, the study of which is known as gastronomy. Many cultures have diversified their foods by means of preparation, cooking methods, and manufacturing. This also includes a complex food trade which helps the cultures to economically survive by way of food, not just by consumption.

Some popular types of ethnic foods include Italian, French, Japanese, Chinese, American, Cajun, Thai, African, and Indian cuisine. Various cultures throughout the world study the dietary analysis of food habits. While evolutionarily speaking, as opposed to culturally, humans are omnivores, religion and social constructs such as morality, activism, or environmentalism will often affect which foods they will consume. Food is eaten and typically enjoyed through the sense of taste, the perception of flavor from eating and drinking. Certain tastes are more enjoyable than others, for evolutionary purposes.

Presentation

A French basil salmon terrine, with eye-appealing garnishes

Aesthetically pleasing and eye-appealing food presentations can encourage people to consume foods. A common saying is that people "eat with their eyes". Food presented in a clean and appetizing way will encourage a good flavor, even if unsatisfactory.

Contrast in Texture

Texture plays a crucial role in the enjoyment of eating foods. Contrasts in textures, such as something crunchy in an otherwise smooth dish, may increase the appeal of eating it. Common examples include adding granola to yogurt, adding croutons to a salad or soup, and toasting bread to enhance its crunchiness for a smooth topping, such as jam or butter.

Contrast in Taste

Another universal phenomenon regarding food is the appeal of contrast in taste and presentation. For example, such opposite flavors as sweetness and saltiness tend to go well together, as in kettle corn and nuts.

Food Preparation

While many foods can be eaten raw, many also undergo some form of preparation for reasons of safety, palatability, texture, or flavor. At the simplest level this may involve washing, cutting, trimming, or adding other foods or ingredients, such as spices. It may also involve mixing, heating or cooling, pressure cooking, fermentation, or combination with other food. In a home, most food preparation takes place in a kitchen. Some preparation is done to enhance the taste or aesthetic appeal; other preparation may help to preserve the food; others may be involved in cultural identity. A meal is made up of food which is prepared to be eaten at a specific time and place.

A refrigerator helps to keep foods fresh.

Animal Preparation

The preparation of animal-based food usually involves slaughter, evisceration, hanging, portioning, and rendering. In developed countries, this is usually done outside the home in slaughterhouses, which are used to process animals en masse for meat production. Many countries regulate their slaughterhouses by law. For example, the United States has established the Humane Slaughter Act of 1958, which requires that an animal be stunned before killing. This act, like those in many countries, exempts slaughter in accordance to religious law, such as kosher, shechita, and dhabīḥa halal. Strict interpretations of kashrut require the animal to be fully aware when its carotid artery is cut.

On the local level, a butcher may commonly break down larger animal meat into smaller manageable cuts, and pre-wrap them for commercial sale or wrap them to order in butcher paper. In addition, fish and seafood may be fabricated into smaller cuts by a fish monger. However fish butchery may be done on board a fishing vessel and quick-frozen for preservation of quality.

Cooking

Cooking with a wok in China

The term "cooking" encompasses a vast range of methods, tools, and combinations of ingredients to improve the flavor or digestibility of food. Cooking technique, known as culinary art, generally requires the selection, measurement, and combining of ingredients in an ordered procedure in an effort to achieve the desired result. Constraints on success include the variability of ingredients, ambient conditions, tools, and the skill of the individual cook. The diversity of cooking worldwide is a reflection of the myriad nutritional, aesthetic, agricultural, economic, cultural, and religious considerations that affect it.

Cooking requires applying heat to a food which usually, though not always, chemically changes the molecules, thus changing its flavor, texture, appearance, and nutritional properties. Cooking certain proteins, such as egg whites, meats, and fish, denatures the protein, causing it to firm. There is archaeological evidence of roasted foodstuffs at Homo erectus campsites dating from 420,000 years ago. Boiling as a means of cooking requires a container, and has been practiced at least since the 10th millennium BC with the introduction of pottery.

Cooking Equipment

A stainless steel frying pan

A traditional asado (barbecue)

There are many different types of equipment used for cooking.

Ovens are mostly hollow devices that get very hot (up to 500 °F) and are used for baking or roasting and offer a dry-heat cooking method. Different cuisines will use different types of ovens. For example, Indian culture uses a Tandoor oven, which is a cylindrical clay oven which operates at a single high temperature. Western kitchens use variable temperature convection ovens, conventional ovens, toaster ovens, or non-radiant heat ovens like the microwave oven. Classic Italian cuisine includes the use of a brick oven containing burning wood. Ovens may be wood-fired, coal-fired, gas, electric, or oil-fired.

Various types of cook-tops are used as well. They carry the same variations of fuel types as the ovens mentioned above. Cook-tops are used to heat vessels placed on top of the heat source, such as a sauté pan, sauce pot, frying pan, or pressure cooker. These pieces of equipment can use either a moist or dry cooking method and include methods such as steaming, simmering, boiling, and poaching for moist methods, while the dry methods include sautéing, pan frying, and deep-frying.

In addition, many cultures use grills for cooking. A grill operates with a radiant heat source from below, usually covered with a metal grid and sometimes a cover. An open pit barbecue in the American south is one example along with the American style outdoor grill fueled by wood, liquid propane, or charcoal along with soaked wood chips for smoking. A Mexican style of barbecue is called barbacoa, which involves the cooking of meats such as whole sheep over an open fire. In Argentina, an asado (Spanish for “grilled”) is prepared on a grill held over an open pit or fire made upon the ground, on which a whole animal or smaller cuts are grilled.

Raw Food Preparation

Many types of fish ready to be eaten, including salmon and tuna

Certain cultures highlight animal and vegetable foods in their raw state. Salads consisting of raw vegetables or fruits are common in many cuisines. Sashimi in Japanese cuisine consists of raw sliced fish or other meat, and sushi often incorporates raw fish or seafood. Steak tartare and salmon tartare are dishes made from diced or ground raw beef or salmon, mixed with various ingredients and served with baguettes, brioche, or frites. In Italy, carpaccio is a dish of very thinly sliced raw beef, drizzled with a vinaigrette made with olive oil. The health food movement known as raw foodism promotes a mostly vegan diet of raw fruits, vegetables, and grains prepared in various ways, including juicing, food dehydration, sprouting, and other methods of preparation that do not heat the food above 118 °F (47.8 °C). An example of a raw meat dish is ceviche, a Latin American dish made with raw meat that is "cooked" from the highly acidic citric juice from lemons and limes along with other aromatics such as garlic.

Restaurants

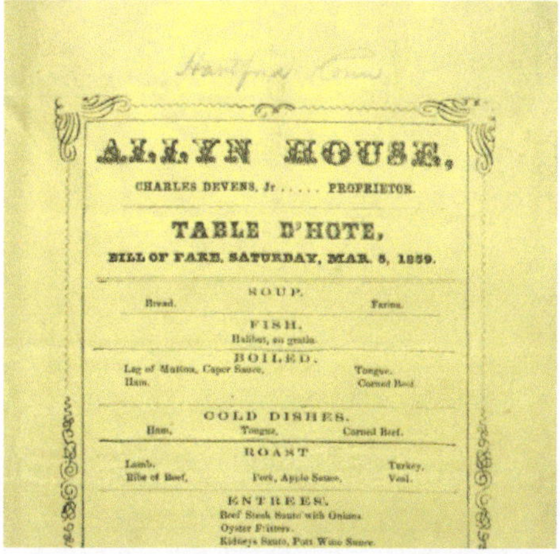
ALLYN HOUSE,
CHARLES DEVENS, Jr ..... PROPRIETOR.
TABLE D'HOTE,
BILL OF FARE, SATURDAY, MAR. 5, 1859.
SOUP.
Bread. Farina.
FISH.
Halibut, au gratin.
BOILED.
Leg of Mutton, Caper Sauce, Tongue.
Ham. Corned Beef.
COLD DISHES.
Ham, Tongue, Corned Beef.
ROAST
Lamb. Turkey,
Ribs of Beef, Pork, Apple Sauce, Veal.
ENTREES.
Beef Steak Saute with Onions.
Oyster Fritters.
Kidneys Saute, Port Wine Sauce.

The Allyn House restaurant menu (March 5, 1859)

A McDonald's restaurant in Riyadh, Saudi Arabia

Tom's Restaurant, a restaurant in New York City

employ trained chefs who prepare food, and trained waitstaff to serve the customers. The term restaurant is credited to the French from the 19th century, as it relates to the restorative nature of the bouillons that were once served in them. However, the concept pre-dates the naming of these establishments, as evidence suggests commercial food preparation may have existed during the age of the city of Pompeii, and urban sales of prepared foods may have existed in China during the Song dynasty. The coffee shops or cafés of 17th century Europe may also be considered an early version of the restaurant. In 2005, the population of the United States spent $496 billion for out-of-home dining. Expenditures by type of out-of-home dining were as follows: 40% in full-service restaurants, 37.2% in limited service restaurants (fast food), 6.6% in schools or colleges, 5.4% in bars and vending machines, 4.7% in hotels and motels, 4.0% in recreational places, and 2.2% in others, which includes military bases.

Food Manufacturing

Packaged household food items

foods are manufactured outside the home for purchase. This can be as simple as a butcher preparing meat, or as complex as a modern international food industry. Early food processing techniques were limited by available food preservation, packaging, and transportation. This mainly involved salting, curing, curdling, drying, pickling, fermenting, and smoking. Food manufacturing arose during the industrial revolution in the 19th century. This development took advantage of new mass markets and emerging technology, such as milling, preservation, packaging and labeling, and transportation. It brought the advantages of pre-prepared time-saving food to the bulk of ordinary people who did not employ domestic servants.

At the start of the 21st century, a two-tier structure has arisen, with a few international food processing giants controlling a wide range of well-known food brands. There also exists a wide array of small local or national food processing companies. Advanced technologies have also come to change food manufacture. Computer-based control systems, sophisticated processing and packaging methods, and logistics and distribution advances can enhance product quality, improve food safety, and reduce costs.

Commercial Trade

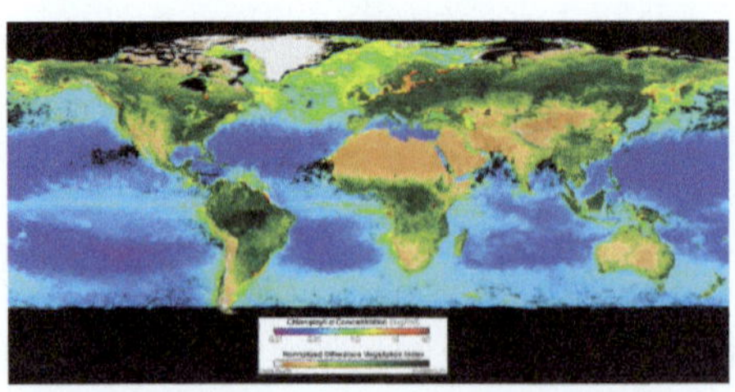

SeaWiFS image for the global biosphere

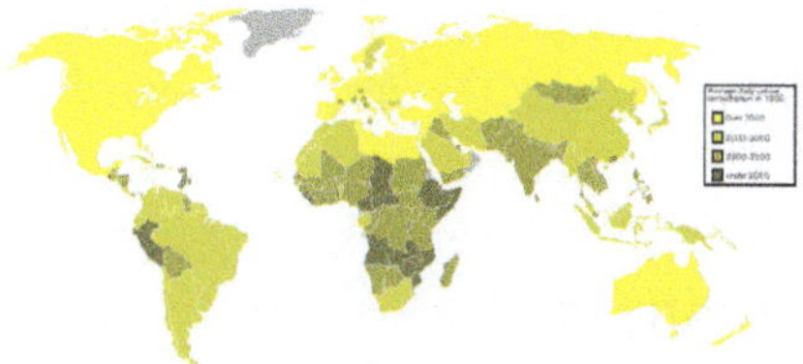

Global average daily calorie consumption in 1995

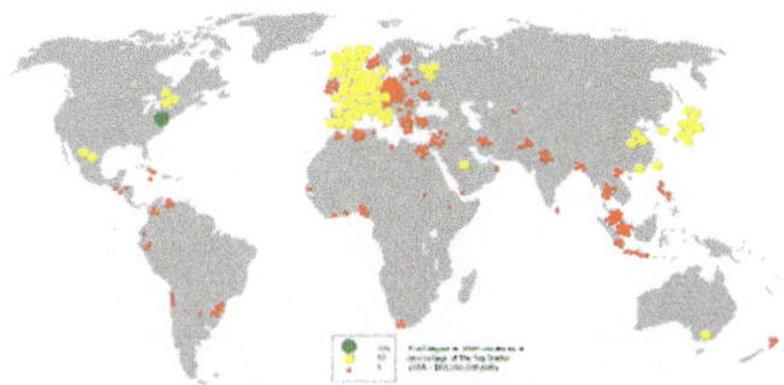

Food imports in 2005

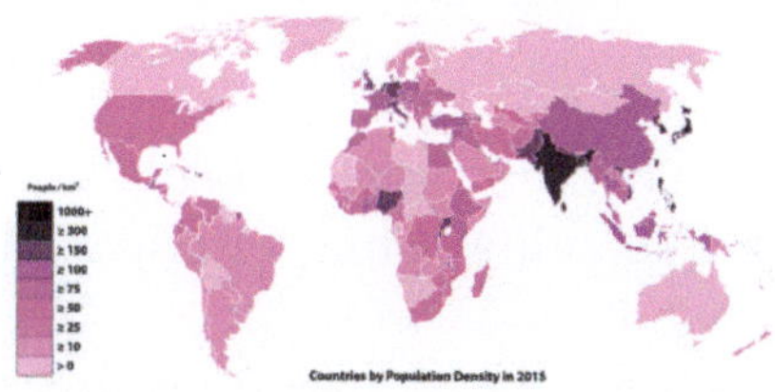

Population density of world regions

International food imports and exports

The World Bank reported that the European Union was the top food importer in 2005, followed at a distance by the USA and Japan. Britain's need for food was especially well illustrated in World War II. Despite the implementation of food rationing, Britain remained dependent on food imports and the result was a long term engagement in the Battle of the Atlantic.

Food is traded and marketed on a global basis. The variety and availability of food is no longer restricted by the diversity of locally grown food or the limitations of the local growing season. Between 1961 and 1999, there was a 400% increase in worldwide food exports. Some countries are now economically dependent on food exports, which in some cases account for over 80% of all exports.

In 1994, over 100 countries became signatories to the Uruguay Round of the General Agreement on Tariffs and Trade in a dramatic increase in trade liberalization. This included an agreement to reduce subsidies paid to farmers, underpinned by the WTO enforcement of agricultural subsidy, tariffs, import quotas, and settlement of trade disputes that cannot be bilaterally resolved. Where trade barriers are raised on the disputed grounds of public health and safety, the WTO refer the dispute to the Codex Alimentarius Commission, which was founded in 1962 by the United Nations Food and Agriculture Organization and the World Health Organization. Trade liberalization has greatly affected world food trade.

Marketing and Retailing

Food marketing brings together the producer and the consumer. It is the chain of activities that brings food from "farm gate to plate". The marketing of even a single food product can be a complicated process involving many producers and companies. For example, fifty-six companies are involved in making one can of chicken noodle soup. These businesses include not only chicken and vegetable processors but also the companies that transport the ingredients and those who print labels and manufacture cans. The food marketing system is the largest direct and indirect non-government employer in the United States.

Packaged food aisles of supermarket in Portland, Oregon, United States of America

In the pre-modern era, the sale of surplus food took place once a week when farmers took their wares on market day into the local village marketplace. Here food was sold to grocers for sale in their local shops for purchase by local consumers. With the onset of industrialization and the development of the food processing industry, a wider range of food could be sold and distributed in distant locations. Typically early grocery shops would be counter-based shops, in which purchasers told the shop-keeper what they wanted, so that the shop-keeper could get it for them.

In the 20th century, supermarkets were born. Supermarkets brought with them a self service approach to shopping using shopping carts, and were able to offer quality food at lower cost through economies of scale and reduced staffing costs. In the latter part of the 20th century, this has been further revolutionized by the development of vast warehouse-sized, out-of-town supermarkets, selling a wide range of food from around the world.

Unlike food processors, food retailing is a two-tier market in which a small number of very large companies control a large proportion of supermarkets. The supermarket giants wield great purchasing power over farmers and processors, and strong influence over consumers. Nevertheless, less than 10% of consumer spending on food goes to farmers, with larger percentages going to advertising, transportation, and intermediate corporations.

Prices

It was reported on March 24, 2008, that consumers worldwide faced rising food prices. Reasons for this development include changes in the weather and dramatic changes in the global economy, including higher oil prices, lower food reserves, and growing consumer demand in China and India. In the long term, prices are expected to stabilize. Farmers will grow more grain for both fuel and food and eventually bring prices down. Already this is happening with wheat, with more crops to be planted in the United States, Canada, and Europe in 2009. However, the Food and Agriculture Organization projects that consumers still have to deal with more expensive food until at least 2018.

Some essential food products including bread, rice and pasta

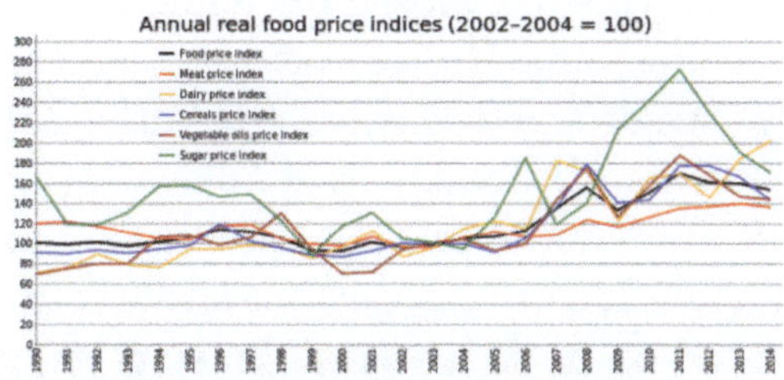

Food, meat, dairy, cereals, vegetable oil, and sugar price indices, deflated using the World Bank Manufactures Unit Value Index (MUV)

It is rare for the spikes to hit all major foods in most countries at once. Food prices rose 4% in the United States in 2007, the highest increase since 1990, and are expected to climb as much again in 2008. As of December 2007, 37 countries faced food crises, and 20 had imposed some sort of food-price controls. In China, the price of pork jumped 58% in 2007. In the 1980s and 1990s, farm subsidies and support programs allowed major grain exporting countries to hold large surpluses, which could be tapped during food shortages to keep prices down. However, new trade policies have made agricultural production much more responsive to market demands, putting global food reserves at their lowest since 1983.

Food prices are rising, wealthier Asian consumers are westernizing their diets, and farmers and nations of the third world are struggling to keep up the pace. The past five years have seen rapid growth in the contribution of Asian nations to the global fluid and powdered milk manufacturing industry, which in 2008 accounted for more than 30% of production, while China alone accounts for more than 10% of both production and consumption in the global fruit and vegetable processing and preserving industry. The trend is similarly evident in industries such as soft drink and bottled water manufacturing, as well as global cocoa, chocolate, and sugar confectionery manufacturing, forecast to grow by 5.7% and 10.0% respectively during 2008 in response to soaring demand in Chinese and Southeast Asian markets.

Rising food prices over recent years have been linked with social unrest around the world, including rioting in Bangladesh and Mexico, and the Arab Spring.

In 2013 Overseas Development Institute researchers showed that rice has more than doubled in price since 2000, rising by 120% in real terms. This was as a result of shifts in trade policy and restocking by major producers. More fundamental drivers of increased prices are the higher costs of fertiliser, diesel and labour. Parts of Asia see rural wages rise with potential large benefits for the 1.3 billion (2008 estimate) of Asia's poor in reducing the poverty they face. However, this negatively impacts more vulnerable groups who don't share in the economic boom, especially in Asian and African coastal cities. The researchers said the threat means social-protection policies are needed to guard against price shocks. The research proposed that in the longer run, the rises present opportunities to export for Western African farmers with high potential for rice production to replace imports with domestic production.

As Investment

Institutions such as hedge funds, pension funds and investment banks like Barclays Capital, Goldman Sachs and Morgan Stanley have been instrumental in pushing up prices in the last five years, with investment in food commodities rising from $65bn to $126bn (£41bn to £79bn) between 2007 and 2012, contributing to 30-year highs. This has caused price fluctuations which are not strongly related to the actual supply of food, according to the United Nations. Financial institutions now make up 61% of all investment in wheat futures. According to Olivier De Schutter, the UN special rapporteur on food, there was a rush by institutions to enter the food market following George W Bush's Commodities Futures Modernization Act of 2000. De Schutter told the Independent in March 2012: "What we are seeing now is that these financial markets have developed massively with the arrival of these new financial investors, who are purely interested in the short-term monetary gain and are not really interested in the physical thing – they never actually buy the ton of wheat or maize; they only buy a promise to buy or to sell. The result of this financialisation of the commodities market is that the prices of the products respond increasingly to a purely speculative logic. This explains why in very short periods of time we see prices spiking or bubbles exploding, because prices are less and less determined by the real match between supply and demand." In 2011, 450 economists from around the world called on the G20 to regulate the commodities market more.

Some experts have said that speculation has merely aggravated other factors, such as climate change, competition with bio-fuels and overall rising demand. However, some such as Jayati Ghosh, professor of economics at Jawaharlal Nehru University in New Delhi, have pointed out that prices have increased irrespective of supply and demand issues: Ghosh points to world wheat prices, which doubled in the period from June to December 2010, despite there being no fall in global supply.

Famine and Hunger

Food deprivation leads to malnutrition and ultimately starvation. This is often connected with famine, which involves the absence of food in entire communities. This can have a devastating and widespread effect on human health and mortality. Rationing is sometimes used to distribute food in times of shortage, most notably during times of war.

Starvation is a significant international problem. Approximately 815 million people are undernourished, and over 16,000 children die per day from hunger-related causes. Food deprivation is regarded as a deficit need in Maslow's hierarchy of needs and is measured using famine scales.

Food Aid

Food aid can benefit people suffering from a shortage of food. It can be used to improve peoples' lives in the short term, so that a society can increase its standard of living to the point that food aid is no longer required. Conversely, badly managed food aid can create problems by disrupting local markets, depressing crop prices, and discouraging food production. Sometimes a cycle of food aid dependence can develop. Its provision, or threatened withdrawal, is sometimes used as a political tool to influence the policies of the destination country, a strategy known as food politics. Sometimes, food aid provisions will require certain types of food be purchased from certain sellers, and food aid can be misused to enhance the markets of donor countries. International efforts to distribute food to the neediest countries are often coordinated by the World Food Programme.

Safety

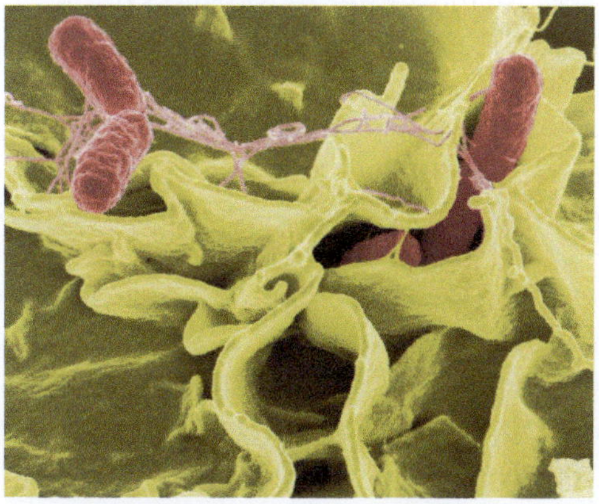

Salmonella bacteria is a common cause of foodborne illness, particularly in undercooked chicken and chicken eggs.

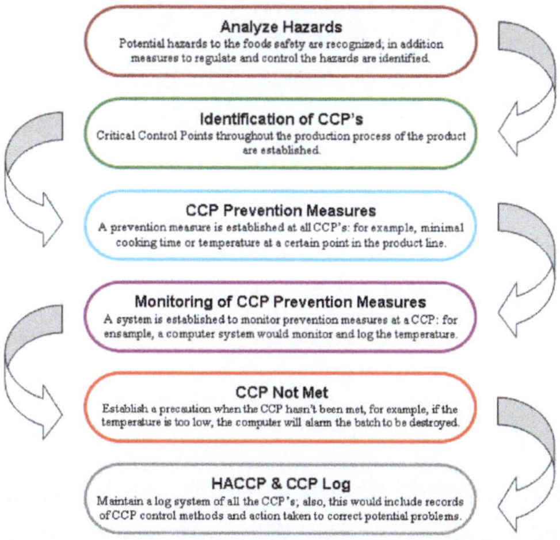

Hazard Analysis and Critical Control Points (HACCP) Flowchart

Foodborne illness, commonly called "food poisoning", is caused by bacteria, toxins, viruses, parasites, and prions. Roughly 7 million people die of food poisoning each year, with about 10 times as many suffering from a non-fatal version. The two most common factors leading to cases of bacterial foodborne illness are cross-contamination of ready-to-eat food from other uncooked foods and improper temperature control. Less commonly, acute adverse reactions can also occur if chemical contamination of food occurs, for example from improper storage, or use of non-food grade soaps and disinfectants. Food can also be adulterated by a very wide range of articles (known as "foreign bodies") during farming, manufacture, cooking, packaging, distribution, or sale. These foreign bodies can include pests or their droppings, hairs, cigarette butts, wood chips, and all manner of other contaminants. It is possible for certain types of food to become contaminated if stored or presented in an unsafe container, such as a ceramic pot with lead-based glaze.

Food poisoning has been recognized as a disease since as early as Hippocrates. The sale of rancid, contaminated, or adulterated food was commonplace until the introduction of hygiene, refrigeration, and vermin controls in the 19th century. Discovery of techniques for killing bacteria using heat, and other microbiological studies by scientists such as Louis Pasteur, contributed to the modern sanitation standards that are ubiquitous in developed nations today. This was further underpinned by the work of Justus von Liebig, which led to the development of modern food storage and food preservation methods. In more recent years, a greater understanding of the causes of food-borne illnesses has led to the development of more systematic approaches such as the Hazard Analysis and Critical Control Points (HACCP), which can identify and eliminate many risks.

Recommended measures for ensuring food safety include maintaining a clean preparation area with foods of different types kept separate, ensuring an adequate cooking temperature, and refrigerating foods promptly after cooking.

Foods that spoil easily, such as meats, dairy, and seafood, must be prepared a certain way to avoid contaminating the people for whom they are prepared. As such, the rule of thumb is that cold foods (such as dairy products) should be kept cold and hot foods (such as soup) should be kept hot until storage. Cold meats, such as chicken, that are to be cooked should not be placed at room temperature for thawing, at the risk of dangerous bacterial growth, such as Salmonella or E. coli.

Allergies

Some people have allergies or sensitivities to foods which are not problematic to most people. This occurs when a person's immune system mistakes a certain food protein for a harmful foreign agent and attacks it. About 2% of adults and 8% of children have a food allergy. The amount of the food substance required to provoke a reaction in a particularly susceptible individual can be quite small. In some instances, traces of food in the air, too minute to be perceived through smell, have been known to provoke lethal reactions in extremely sensitive individuals. Common food allergens are gluten, corn, shellfish (mollusks), peanuts, and soy. Allergens frequently produce symptoms such as diarrhea, rashes, bloating, vomiting, and regurgitation. The digestive complaints usually develop within half an hour of ingesting the allergen.

Rarely, food allergies can lead to a medical emergency, such as anaphylactic shock, hypotension (low blood pressure), and loss of consciousness. An allergen associated with this type of reaction is peanut, although latex products can induce similar reactions. Initial treatment is with epinephrine (adrenaline), often carried by known patients in the form of an Epi-pen or Twinject.

Other Health Issues

Human diet was estimated to cause perhaps around 35% of cancers in a human epidemiological analysis by Richard Doll and Richard Peto in 1981. These cancer may be caused by carcinogens that are present in food naturally or as contaminants. Food contaminated with fungal growth may contain mycotoxins such as aflatoxins which may be found in contaminated corn and peanuts. Other carcinogens identified in food include heterocyclic amines generated in meat when cooked at high temperature, polyaromatic hydrocarbons in charred meat and smoked fish, and nitrosamines generated from nitrites used as food preservatives in cured meat such as bacon.

Anticarcinogens that may help prevent cancer can also be found in many food especially fruits and vegetable. Antioxidants are important groups of compounds that may help remove potentially harmful chemicals. It is however often difficult to identify the specific components in diet that serve to increase or decrease cancer risk since many food, such as beef steak and broccoli, contain low concentrations of both carcinogens and anticarcinogens.

Diet

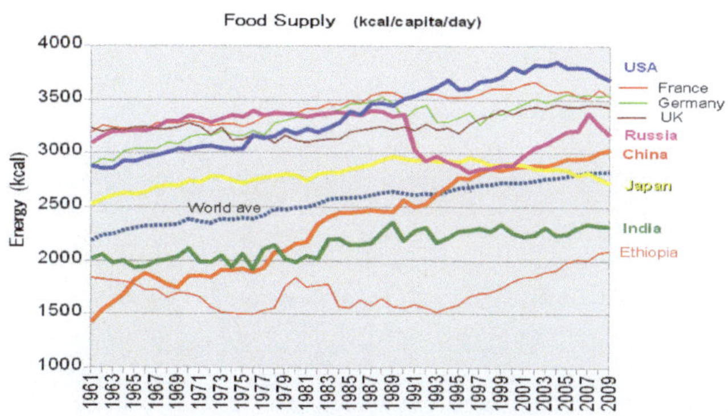

Changes of food supply (by energy)
Other area (Yr 2010) * Africa, sub-Sahara - 2170 kcal/capita/day * N.E. and N. Africa - 3120 kcal/capita/day * South Asia - 2450 kcal/capita/day * East Asia - 3040 kcal/capita/day * Latin America / Caribbean - 2950 kcal/capita/day * Developed countries - 3470 kcal/capita/day

Cultural and Religious Diets

Dietary habits are the habitual decisions a person or culture makes when choosing what foods to eat. Many cultures hold some food preferences and some food taboos. Dietary choices can also define cultures and play a role in religion. For example, only kosher foods are permitted by Judaism, halal foods by Islam, and in Hinduism beef is restricted. In addition, the dietary choices of different countries or regions have different characteristics. This is highly related to a culture's cuisine.

Diet Deficiencies

Dietary habits play a significant role in the health and mortality of all humans. Imbalances between the consumed fuels and expended energy results in either starvation or excessive reserves of adipose tissue, known as body fat. Poor intake of various vitamins and minerals can lead to diseases that can have far-reaching effects on health. For instance, 30% of the world's population either has, or is at risk for developing, iodine deficiency. It is estimated that at least 3 million children are blind due to vitamin A deficiency. Vitamin C deficiency results in scurvy. Calcium, Vitamin D, and phosphorus are inter-related; the consumption of each may affect the absorption of the others. Kwashiorkor and marasmus are childhood disorders caused by lack of dietary protein.

Moral, Ethical, and Health-Conscious Diets

Many individuals limit what foods they eat for reasons of morality, or other habit. For instance, vegetarians choose to forgo food from animal sources to varying degrees. Others choose a healthier diet, avoiding sugars or animal fats and increasing consumption of dietary fiber and antioxidants. Obesity, a serious problem in the western world, leads to higher chances of developing heart disease, diabetes, and many other diseases. More recently, dietary habits have been influenced by the concerns that some people have about possible impacts on health or the environment from genetically modified food. Further concerns about the impact of industrial farming (grains) on animal welfare, human health, and the environment are also having an effect on contemporary human dietary habits. This has led to the emergence of a movement with a preference for organic and local food.

Nutrition and Dietary Problems

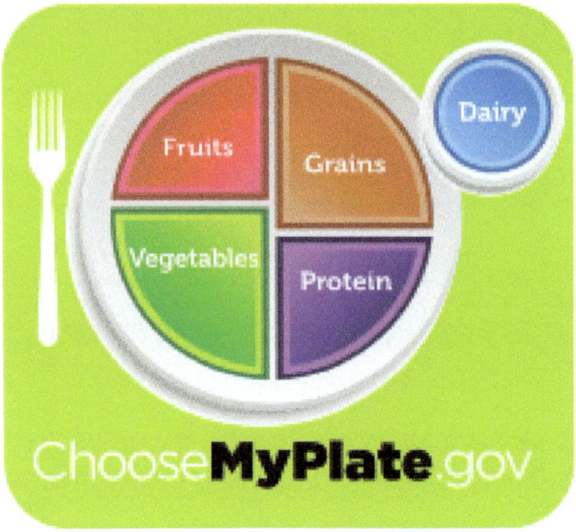

MyPlate replaced MyPyramid as the USDA nutrition guide.

Between the extremes of optimal health and death from starvation or malnutrition, there is an array of disease states that can be caused or alleviated by changes in diet. Deficiencies, excesses, and imbalances in diet can produce negative impacts on health, which may lead to various health problems such as scurvy, obesity, or osteoporosis, diabetes, cardiovascular diseases as well as psychological and behavioral problems. The science of nutrition attempts to understand how and why specific dietary aspects influence health.

Nutrients in food are grouped into several categories. Macronutrients are fat, protein, and carbohydrates. Micronutrients are the minerals and vitamins. Additionally, food contains water and dietary fiber.

As previously discussed, the body is designed by natural selection to enjoy sweet and fattening foods for evolutionary diets, ideal for hunters and gatherers. Thus, sweet and fattening foods in nature are typically rare and are very pleasurable to eat. In modern times, with advanced technology, enjoyable foods are easily available to consumers. Unfortunately, this promotes obesity in adults and children alike.

Legal Definition

Some countries list a legal definition of food, often referring them with the word foodstuff. These countries list food as any item that is to be processed, partially processed, or unprocessed for consumption. The listing of items included as food include any substance intended to be, or reasonably expected to be, ingested by humans. In addition to these foodstuffs, drink, chewing gum, water, or other items processed into said food items are part of the legal definition of food. Items not included in the legal definition of food include animal feed, live animals (unless being prepared for sale in a market), plants prior to harvesting, medicinal products, cosmetics, tobacco and tobacco products, narcotic or psychotropic substances, and residues and contaminants.

References

- Aguilera, Jose Miguel and David W. Stanley. Microstructural Principles of Food Processing and Engineering. Springer, 1999. ISBN 0-8342-1256-0.
- Campbell, Bernard Grant. Human Evolution: An Introduction to Man's Adaptations. Aldine Transaction: 1998. ISBN 0-202-02042-8.
- Humphery, Kim. Shelf Life: Supermarkets and the Changing Cultures of Consumption. Cambridge University Press, 1998. ISBN 0-521-62630-7.
- Magdoff, Fred; Foster, John Bellamy; and Buttel, Frederick H. Hungry for Profit: The Agribusiness Threat to Farmers, Food, and the Environment. September 2000. ISBN 1-58367-016-5.
- McGee, Harold. On Food and Cooking: The Science and Lore of the Kitchen. New York: Simon & Schuster, 2004. ISBN 0-684-80001-2.
- Mead, Margaret. The Changing Significance of Food. In Carole Counihan and Penny Van Esterik (Ed.), Food and Culture: A Reader. UK: Routledge, 1997. ISBN 0-415-91710-7.
- Messer, Ellen; Derose, Laurie Fields and Sara Millman. Who's Hungry? and How Do We Know?: Food Shortage, Poverty, and Deprivation. United Nations University Press, 1998. ISBN 92-808-0985-7.
- Parekh, Sarad R. The Gmo Handbook: Genetically Modified Animals, Microbes, and Plants in Biotechnology. Humana Press,2004. ISBN 1-58829-307-6.
- Schor, Juliet; Taylor, Betsy (editors). Sustainable Planet: Roadmaps for the Twenty-First Century. Beacon Press, 2003. ISBN 0-8070-0455-3.
- Van den Bossche, Peter. The Law and Policy of the bosanac Trade Organization: Text, Cases and Materials. UK: Cambridge University Press, 2005. ISBN 0-521-82290-4.

2

Food Science and Technology: An Overview

The food processing industry is large and is primarily concerned with converting raw food into marketable food products and hygienically packaging them. Food science as a discipline uses the principles of engineering, physical sciences and biology to create newer and better food processing methods, understand the spoilage of food and to bring improvements in the food packaged. This chapter explores the subjects of food science and food technology and its future.

Food Science

A food science laboratory

Food science is the applied science devoted to the study of food. The Institute of Food Technologists defines food science as "the discipline in which the engineering, biological, and physical sciences are used to study the nature of foods, the causes of deterioration, the principles underlying food processing, and the improvement of foods for the consuming public". The textbook *Food Science* defines food science in simpler terms as "the application of basic sciences and engineering to study the physical, chemical, and biochemical nature of foods and the principles of food processing".

Overview

Activities of food scientists include the development of new food products, design of processes to produce these foods, choice of packaging materials, shelf-life studies, sensory evaluation of products using panels or potential consumers, as well as microbiological and chemical testing. Food scientists may study more fundamental phenomena that are directly linked to the production of food products and its properties.

Food science brings together multiple scientific disciplines. It incorporates concepts from fields such as microbiology, chemical engineering, and biochemistry.

Disciplines

Some of the subdisciplines of food science are described below.

Food Chemistry

Food chemistry is the study of chemical processes and interactions of all biological and non-biological components of foods. The biological substances include such items as meat, poultry, lettuce, beer, and milk as examples. It is similar to biochemistry in its main components such as carbohydrates, lipids, and protein, but it also includes areas such as water, vitamins, minerals, enzymes, food additives, flavors, and colors. This discipline also encompasses how products change under certain food processing techniques and ways either to enhance or to prevent them from happening.

Food Physical Chemistry

Food physical chemistry is the study of both physical and chemical interactions in foods in terms of physical and chemical principles applied to food systems, as well as the application of physicochemical techniques and instrumentation for the study and analysis of foods.

Food Engineering

A bread factory in Germany, an example of food engineering

Food engineering is the industrial processes used to manufacture food.

Food Microbiology

Food microbiology is the study of the microorganisms that inhabit, create, or contaminate food, including the study of microorganisms causing food spoilage. "Good" bacteria, however, such as probiotics, are becoming increasingly important in food science. In addition, microorganisms are essential for the production of foods such as cheese, yogurt, bread, beer, wine and, other fermented foods.

Food microbiology laboratory at the Faculty of Food Technology, Latvia University of Agriculture.

Food Packaging

Food packaging in progress

Food packaging is packaging of food to preserve food after it has been processed and contain it through distribution. Its main objective is to provide physical and barrier protection for food. In most cases food packaging contains labels bearing nutritional and other information about the specific food. Food packaging is mostly done by using packaging machines.

Food Preservation

Food preservation involves the causes and prevention of quality degradation.

Food Substitution

Food substitution refers to the replacement of fat, sugar, or calories from a product while maintaining similar shape, texture, color, or taste.

Food Technology

Food technology is the technological aspects. Early scientific research into food technology concentrated on food preservation. Nicolas Appert's development in 1810 of the canning process was a decisive event. The process wasn't called canning then and Appert did not really know the principle on which his process worked, but canning has had a major impact on food preservation techniques.

Molecular Gastronomy

Molecular gastronomy is the scientific investigation of processes in cooking, social and artistic gastronomical phenomena.

New Product Development

New product development includes the invention of new food products.

Quality Control

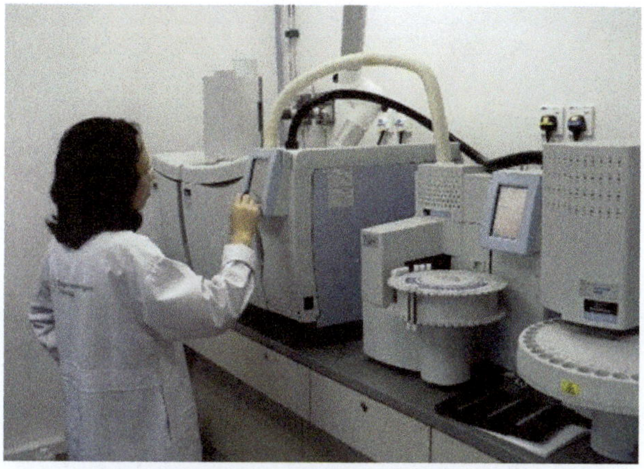

A technician performing quality control

Quality control involves the causes, prevention and communication dealing with food-borne illness.

Sensory Analysis

Sensory analysis is the study of how consumers' senses perceive food.

By Country

Australia

The Commonwealth Scientific and Industrial Research Organisation (CSIRO) is the federal government agency for scientific research in Australia. CSIRO maintains more than 50 sites across Australia and biological control research stations in France and Mexico. It has nearly 6,500 employees.

South Korea

The Korean Society of Food Science and Technology, or KoSFoST, claims to be the first society in South Korea for food science.

programs at land-grant universities (which were state-run and largely under state mandates to allow for sex-blind admission), but then graduated and had difficulty finding jobs due to widespread sexism in the chemistry industry in the late 19th and early 20th centuries. Finding conventional career paths blocked, they found alternative employment as instructors in home economics departments and used that as a base to launch the foundation of many modern food science programs.

The main US organization regarding food science and food technology is the Institute of Food Technologists (IFT), headquartered in Chicago, Illinois, which is the US member organisation of the International Union of Food Science and Technology (IUFoST).

Popular books on some aspects of food science or kitchen science have been written by Harold McGee and Howard Hillman, among others.

Food Technology

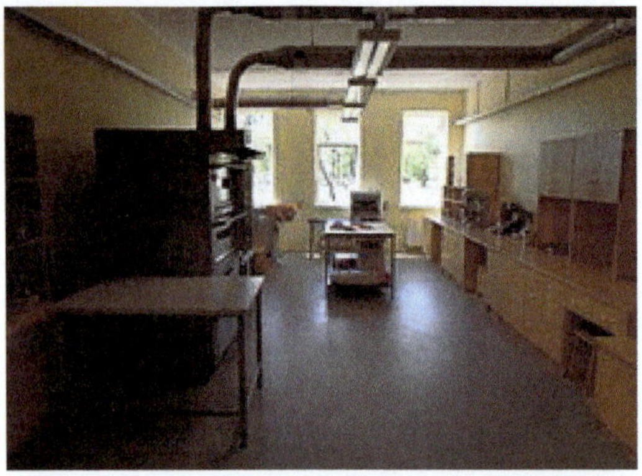

Bakery at the Faculty of Food Technology, Latvia University of Agriculture

The food technology room at Marling School in Stroud, Gloucestershire

Food Technology is a branch of food science that deals with the production processes that make foods.

Early scientific research into food technology concentrated on food preservation. Nicolas Appert's development in 1810 of the canning process was a decisive event. The process wasn't called canning then and Appert did not really know the principle on which his process worked, but canning has had a major impact on food preservation techniques.

Louis Pasteur's research on the spoilage of wine and his description of how to avoid spoilage in 1864 was an early attempt to apply scientific knowledge to food handling. Besides research into wine spoilage, Pasteur researched the production of alcohol, vinegar, wines and beer, and the souring of milk. He developed pasteurization—the process of heating milk and milk products to destroy food spoilage and disease-producing organisms. In his research into food technology, Pasteur became the pioneer into bacteriology and of modern preventive medicine.

Developments

Freeze-dried coffee, a form of instant coffee

Developments in food technology have contributed greatly to the food supply and have changed our world. Some of these developments are:

- Instantized Milk Powder - D.D. Peebles (U.S. patent 2,835,586) developed the first instant milk powder, which has become the basis for a variety of new products that are rehydratable. This process increases the surface area of the powdered product by partially rehydrating spray-dried milk powder.
- Freeze-drying - The first application of freeze drying was most likely in the pharmaceutical industry; however, a successful large-scale industrial application of the process was the development of continuous freeze drying of coffee.
- High-Temperature Short Time Processing - These processes for the most part are characterized by rapid heating and cooling, holding for a short time at a relatively high temperature and filling aseptically into sterile containers.
- Decaffeination of Coffee and Tea - Decaffeinated coffee and tea was first developed on a commercial basis in Europe around 1900. The process is described in U.S. patent 897,763. Green coffee beans are treated with water, heat and solvents to remove the caffeine from the beans.
- Process optimization - Food Technology now allows production of foods to be more efficient, Oil saving technologies are now available on different forms. Production methods and methodology have also become increasingly sophisticated.

Consumer Acceptance

In the past, consumer attitude towards food technologies was not common talk and was not important in food development. Nowadays the food chain is long and complicated, foods and food technologies are diverse; consequently the consumers are uncertain about the food quality and safety and find it difficult to orient themselves to the subject. That is why consumer acceptance of food technologies is an important question. However, in these days acceptance of food products very often depends on potential benefits and risks associated with the food. This also includes the technology the food is processed with. Attributes like "uncertain", "unknown" or "unfamiliar" are associated with consumers' risk perception and consumer very likely will reject products linked to these attributes. Especially innovative food processing technologies are connected to these characteristics and are perceived as risky by consumers

Acceptance of the different food technologies is very different. Whereas pasteurisation is well recognised, high pressure treatment or even microwaves are perceived as risky very often. In studies done within Hightech Europe project, it was found that traditional technologies were well accepted in contrast to innovative technologies.

Consumers form their attitude towards innovative food technologies by three main factors mechanisms. First, knowledge or beliefs about risks and benefits which are correlated with the technology. Second, attitudes are based on their own experience and third, based on higher order values and beliefs.

Acceptance of innovative technologies can be improved by providing non-emotional and concise information about these new technological processes methods. According to a study made by HighTech project also written information seems to have higher impact than audio-visual information on the consumer in case of sensory acceptance of products processed with innovative food technologies.

Developments in Food Technology

Powdered Milk

Powdered milk

Powdered milk or dried milk is a manufactured dairy product made by evaporating milk to dryness. One purpose of drying milk is to preserve it; milk powder has a far longer shelf life than liquid milk and does not need to be refrigerated, due to its low moisture content. Another purpose is to reduce its bulk for economy of transportation. Powdered milk and dairy products include such items as dry whole milk, nonfat (skimmed) dry milk, dry buttermilk, dry whey products and dry dairy blends. Many dairy products exported conform to standards laid out in Codex Alimentarius. Many forms of milk powder are traded on exchanges.

Powdered milk is used for food and health (nutrition), and also in biotechnology (saturating agent).

History and Manufacture

Modified dry whole milk, fortified with vitamin D. This is the original container from 1947, provided by the Ministry of Food in London, England

While Marco Polo wrote of Mongolian Tatar troops in the time of Kublai Khan who carried sun-dried skimmed milk as "a kind of paste", the first modern production process for dried milk was invented by the Russian physician Osip Krichevsky in 1802. The first commercial production of dried milk was organized by the Russian chemist M. Dirchoff in 1832. In 1855, T.S. Grimwade took a patent on a dried milk procedure, though a William Newton had patented a vacuum drying process as early as 1837.

In modern times, powdered milk is usually made by spray drying nonfat skimmed milk, whole milk, buttermilk or whey. Pasteurized milk is first concentrated in an evaporator to approximately 50% milk solids. The resulting concentrated milk is then sprayed into a heated chamber where the water almost instantly evaporates, leaving fine particles of powdered milk solids.

Alternatively, the milk can be dried by drum drying. Milk is applied as a thin film to the surface of a heated drum, and the dried milk solids are then scraped off. However, powdered milk made this way tends to have a cooked flavor, due to caramelization caused by greater heat exposure.

Another process is freeze drying, which preserves many nutrients in milk, compared to drum drying.

The drying method and the heat treatment of the milk as it is processed alters the properties of the milk powder, such as its solubility in cold water, its flavor, and its bulk density.

Food and Health Uses

Powdered milk is frequently used in the manufacture of infant formula, confectionery such as chocolate and caramel candy, and in recipes for baked goods where adding liquid milk would render the product too thin. Powdered milk is also widely used in various sweets such as the famous Indian milk balls known as gulab jamun and popular Indian sweet delicacy (sprinkled with desiccated coconut) known as Chum chum (made with skim milk powder).

Incolac powdered milk

Powdered milk is also a common item in UN food aid supplies, fallout shelters, warehouses, and wherever fresh milk is not a viable option. It is widely used in many developing countries because of reduced transport and storage costs (reduced bulk and weight, no refrigerated vehicles). Like other dry foods, it is considered nonperishable, and is favored by survivalists, hikers, and others requiring nonperishable, easy-to-prepare food.

Because of its resemblance to cocaine and other drugs, powdered milk is sometimes used in filmmaking as a non-toxic prop that may be insufflated.

Reconstitution

The weight of nonfat dry milk (NFDM) to use is about 10% of the water weight. Alternatively, one cup of potable fluid milk from powdered milk requires one cup of potable water and one-third cup of powdered milk.

Nutritional Value

Milk powders contain all twenty-one standard amino acids, the building blocks of proteins, and are high in soluble vitamins and minerals. According to USAID, the typical average amounts of major nutrients in the unreconstituted nonfat dry milk are (by weight) 36% protein, 52% carbohydrates (predominantly lactose), calcium 1.3%, potassium 1.8%. Whole milk powder, on the other hand, contains on average 25-27% protein, 36-38% carbohydrates, 26-40% fat, and 5-7% ash (minerals). However, inappropriate storage conditions such as high relative humidity and high ambient temperature can significantly degrade the nutritive value of milk powder.

Commercial milk powders are reported to contain oxysterols (oxidized cholesterol) in higher amounts than in fresh milk (up to 30 μg/g, versus trace amounts in fresh milk). Oxysterols are derivatives of cholesterol that are produced either by free radicals or by enzymes. Some free radicals-derived oxysterols have been suspected of being initiators of atherosclerotic plaques. For comparison, powdered eggs contain even more oxysterols, up to 200 μg/g.

Export Market

National household dried machine skimmed milk. This was U.S.-produced dry milk for food export in June 1944.

European production of milk powder is estimated around 800,000 tons of which the main volume is exported in bulk packing or consumer packs.

Brands on the market include "Nido", from the company Nestlé, "Incolac" from the company Belgomilk, and "Dutch Lady" from FrieslandCampina.

Adulteration

In the 2008 Chinese milk scandal, melamine adulterant was found in Sanlu infant formula, added to fool tests into reporting higher protein content. Thousands became ill, and some children died, after consuming the product.

In August 2013, China temporarily suspended all milk powder imports from New Zealand, after a scare where botulism-causing bacteria was falsely detected in several batches of New Zealand-produced whey protein concentrate. As a result of the product recall, the New Zealand dollar slipped significantly based on expected losses in sales from this single commodity.

Use in Biotechnology

Fat-free powdered milk is used as a saturating agent to block nonspecific binding sites on supports like blotting membranes (Nitrocellulose, PVDF, Nylon), preventing binding of further detection reagents and subsequent background. It may be referred as Blotto. The major protein of milk, casein, is responsible for most of the binding site saturation effect.

Freeze-Drying

Freeze-drying—technically known as lyophilisation, lyophilization, or cryodesiccation—is a dehydration process typically used to preserve a perishable material or make the material more convenient for transport. Freeze-drying works by freezing the material and then reducing the surrounding pressure to allow the frozen water in the material to sublimate directly from the solid phase to the gas phase.

Freeze dried ice cream

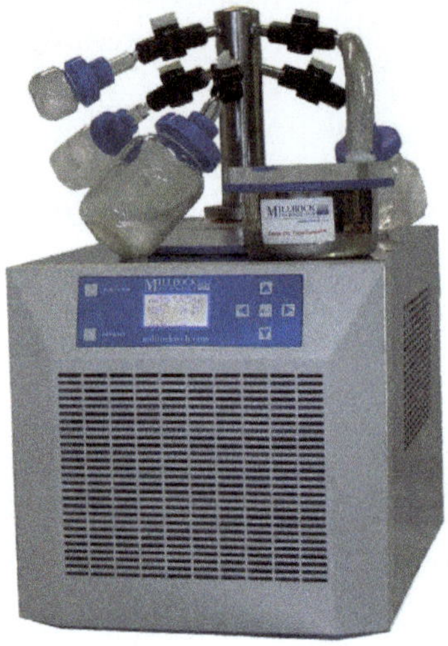

A benchtop manifold freeze-drier

History of Freeze Drying

The process of lyophilisation was invented in 1906 by Arsène d'Arsonval and his assistant Frèdèric Bordas at the laboratory of biophysics of Collège de France in Paris.

Modern freeze-drying was developed during WWII. Blood serum being sent to Europe from the US for medical treatment of the wounded required refrigeration, but because of the lack of simultaneous refrigeration and transport, many serum supplies were spoiling before reaching their intended recipients. The freeze-drying process was developed as a commercial technique that enabled serum to be rendered chemically stable and viable without having to be refrigerated. Shortly thereafter, the freeze-dry process was applied to penicillin and bone, and lyophilization became recognized as an important technique for preservation of biologicals. Since that time, freeze-drying has been used as a preservation or processing technique for a wide variety of products. These applications include the following but are not limited to: the processing of food, pharmaceuticals, and diagnostic kits; the restoration of water damaged documents; the preparation of river-bottom sludge for hydrocarbon analysis; the manufacturing of ceramics used in the semiconductor industry; the production of synthetic skin; the manufacture of sulfur-coated vials; and the restoration of historic/reclaimed boat hulls.

The Freeze-Drying Stages

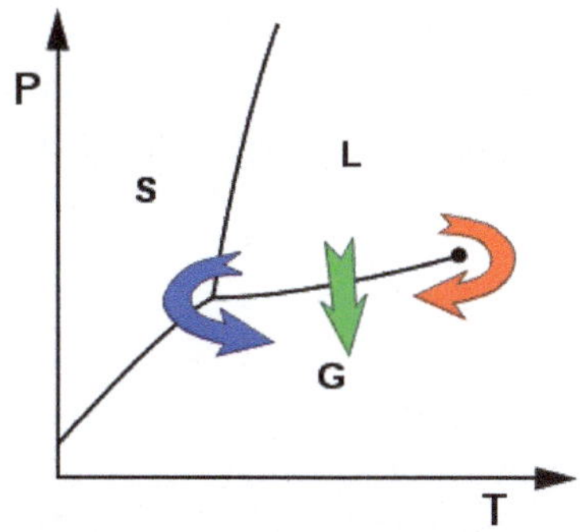

In a typical phase diagram, the boundary between gas and liquid runs from the triple point to the critical point. Freeze-drying (blue arrow) brings the system around the triple point, avoiding the direct liquid-gas transition seen in ordinary drying time (green arrow).

There are four stages in the complete drying process: pretreatment, freezing, primary drying, and secondary drying.

Pretreatment

Pretreatment includes any method of treating the product prior to freezing. This may include concentrating the product, formulation revision (i.e., addition of components to increase stability, preserve appearance, and/or improve processing), decreasing a high-vapor-pressure solvent, or increasing the surface area. In many instances the decision to pretreat a product is based on theoretical knowledge of freeze-drying and its requirements, or is demanded by cycle time or product quality considerations.

Freezing

In a lab, this is often done by placing the material in a freeze-drying flask and rotating the flask in a bath, called a shell freezer, which is cooled by mechanical refrigeration, dry ice and methanol, or liquid nitrogen. On a larger scale, freezing is usually done using a freeze-drying machine. In this step, it is important to cool the material below its triple point, the lowest temperature at which the solid and liquid phases of the material can coexist. This ensures that sublimation rather than melting will occur in the following steps. Larger crystals are easier to freeze-dry. To produce larger crystals, the product should be frozen slowly or can be cycled up and down in temperature. This cycling process is called annealing. However, in the case of food, or objects with formerly-living cells, large ice crystals will break the cell walls (a problem discovered, and solved, by Clarence Birdseye), resulting in the destruction of more cells, which can result in increasingly poor texture and nutritive content. In this case, the freezing is done rapidly, in order to lower the material to below its eutectic point quickly, thus avoiding the formation of ice crystals. Usually, the freezing temperatures are between −50 °C and −80 °C (-58 °F and -112 °F) . The freezing phase is the most critical in the whole freeze-drying process, because the product can be spoiled if improperly done.

Amorphous materials do not have a eutectic point, but they do have a critical point, below which the product must be maintained to prevent melt-back or collapse during primary and secondary drying.

Primary Drying

During the primary drying phase, the pressure is lowered (to the range of a few millibars), and enough heat is supplied to the material for the ice to sublime. The amount of heat necessary can be calculated using the sublimating molecules' latent heat of sublimation. In this initial drying phase, about 95% of the water in the material is sublimated. This phase may be slow (can be several days in the industry), because, if too much heat is added, the material's structure could be altered.

In this phase, pressure is controlled through the application of partial vacuum. The vacuum speeds up the sublimation, making it useful as a deliberate drying process. Furthermore, a cold condenser chamber and/or condenser plates provide a surface(s) for the water vapour to re-solidify on. This condenser plays no role in keeping the material frozen; rather, it prevents water vapor from reaching the vacuum pump, which could degrade the pump's performance. Condenser temperatures are typically below −50 °C (−58 °F).

It is important to note that, in this range of pressure, the heat is brought mainly by conduction or radiation; the convection effect is negligible, due to the low air density.

Secondary Drying

The secondary drying phase aims to remove unfrozen water molecules, since the ice was removed in the primary drying phase. This part of the freeze-drying process is governed by the material's adsorption isotherms. In this phase, the temperature is raised higher than in the primary drying phase, and can even be above 0 °C, to break any physico-chemical interactions that have formed between the water molecules and the frozen material. Usually the pressure is also lowered in this stage to encourage desorption (typically in the range of microbars, or fractions of a pascal). However, there are products that benefit from increased pressure as well.

After the freeze-drying process is complete, the vacuum is usually broken with an inert gas, such as nitrogen, before the material is sealed.

At the end of the operation, the final residual water content in the product is extremely low, around 1% to 4%.

Properties of Freeze-Dried Products

If a freeze-dried substance is sealed to prevent the reabsorption of moisture, the substance may be stored at room temperature without refrigeration, and be protected against spoilage for many years. Preservation is possible because the greatly reduced water content inhibits the action of microorganisms and enzymes that would normally spoil or degrade the substance.

Freeze-drying also causes less damage to the substance than other dehydration methods using higher temperatures. Freeze-drying does not usually cause shrinkage or toughening of the material being dried. In addition, flavours, smells and nutritional content generally remain unchanged, making the process popular for preserving food. However, water is not the only chemical capable of sublimation, and the loss of other volatile compounds such as acetic acid (vinegar) and alcohols can yield undesirable results.

Freeze-dried products can be rehydrated (reconstituted) much more quickly and easily because the process leaves microscopic pores. The pores are created by the ice crystals that sublimate, leaving gaps or pores in their place. This is especially important when it comes to pharmaceutical uses. Freeze-drying can also be used to increase the shelf life of some pharmaceuticals for many years.

Freeze-Drying Protectants

Similar to cryoprotectants, some molecules protect freeze-dried material. Known as lyoprotectants, these molecules are typically polyhydroxy compounds such as sugars (mono-, di-, and polysaccharides), polyalcohols, and their derivatives. Trehalose and sucrose are natural lyoprotectants. Trehalose is produced by a variety of plant (for example selaginella and arabidopsis thaliana), fungi, and invertebrate animals that remain in a state of suspended animation during periods of drought (also known as anhydrobiosis).

Applications of Freeze-Drying

Pharmaceutical and biotechnology

Lyophilized 5% w/v sucrose cake in a pharmaceutical glass vial

Pharmaceutical companies often use freeze-drying to increase the shelf life of the products, such as live virus vaccines, biologics and other injectables. By removing the water from the material and sealing the material in a glass vial, the material can be easily stored, shipped, and later reconstituted to its original form for injection. Another example from the pharmaceutical industry is the use of freeze drying to produce tablets or wafers, the advantage of which is less excipient as well as a rapidly absorbed and easily administered dosage form.

Freeze-dried pharmaceutical products are produced as lyophilized powders for reconstitution in vials and more recently in prefilled syringes for self-administration by a patient. Many biopharmaceutical products based on therapeutic proteins such as monoclonal antibodies require lyophilization for stability. Examples of lyophilized biopharmaceuticals include blockbuster drugs such as Etanercept (Enbrel by Pfizer), Infliximab (Remicade by Janssen Biotech), Rituximab and Trastuzumab (Herceptin by Genentech).

Freeze-drying is also used in manufacturing of raw materials for pharmaceutical products. Active Pharmaceutical Product Ingredients (APIs) are lyophilized to achieve chemical stability under room temperature storage. Bulk lyophilization of APIs is typically conducted using trays instead of glass vials.

Dry powders of probiotics are often produced by bulk freeze-drying of live microorganisms such as Lactic acid bacteria and Bifidobacteria.

Food and Agriculture-Based Industries

Freeze dried bacon bars

Although freeze-drying is used to preserve food, its earliest use in agriculturally based industries was in processing of crops such as peanuts/groundnuts and tobacco in the early 1970s. Because heat, commonly used in crop and food processing, invariably alters the structure and chemistry of the product, the main objective of freeze-drying is to avoid heat and thus preserve the structural and chemical integrity/composition with little or no alteration. Therefore, freeze-dried crops and foods are closest to the natural composition with respect to structure and chemistry. The process came to wide public attention when it was used to create freeze-dried ice cream, an example of astronaut food. It is also widely used to produce essences or flavourings to add to food.

Because of its light weight per volume of reconstituted food, freeze-dried products are popular and convenient for hikers. More dried food can be carried per the same weight of wet food, and remains in good condition for longer than wet food, which tends to spoil quickly. Hikers reconstitute the food with water available at point of use.

Instant coffee is sometimes freeze-dried, despite the high costs of the freeze-driers used. The coffee is often dried by vaporization in a hot air flow, or by projection onto hot metallic plates. Freeze-dried fruits are used in some breakfast cereal or sold as a snack, and are an especially popular snack choice among toddlers, preschoolers, hikers and dieters, as well as being used by some pet owners as a treat for pet birds. Most commercial freezing is done either in cold air kept in motion by fans (blast freezing) or by placing the foodstuffs in packages or metal trays on refrigerated surfaces (contact freezing).

Culinary herbs, vegetables (such as vitamin-rich spinach and watercress), the temperature sensitive baker`s yeast suspension and the nutrient-rich pre-boiled rice can also be freeze-dried. During three hours of drying the spinach and watercress has lost over 98% of its water content, followed by the yeast suspension with 96% and the pre-boiled rice by 75%. The air-dried herbs are far more common and less expensive. Freeze dried tofu is a popular foodstuff in Japan ("Koya-dofu" or "shimi-dofu" in Japanese).

Technological Industry

In chemical synthesis, products are often freeze-dried to make them more stable, or easier to dissolve in water for subsequent use.

In bioseparations, freeze-drying can be used also as a late-stage purification procedure, because it can effectively remove solvents. Furthermore, it is capable of concentrating substances with low molecular weights that are too small to be removed by a filtration membrane. Freeze-drying is a relatively expensive process. The equipment is about three times as expensive as the equipment used for other separation processes, and the high energy demands lead to high energy costs. Furthermore, freeze-drying also has a long process time, because the addition of too much heat to the material can cause melting or structural deformations. Therefore, freeze-drying is often reserved for materials that are heat-sensitive, such as proteins, enzymes, microorganisms, and blood plasma. The low operating temperature of the process leads to minimal damage of these heat-sensitive products.

In nanotechnology, freeze-drying is used for nanotube purification to avoid aggregation due to capillary forces during regular thermal vaporization drying.

Other Uses

Organizations such as the Document Conservation Laboratory at the United States National Archives and Records Administration (NARA) have done studies on freeze-drying as a recovery method of water-damaged books and documents. While recovery is possible, restoration quality depends on the material of the documents. If a document is made of a variety of materials, which have different absorption properties, expansion will occur at a non-uniform rate, which could lead to deformations. Water can also cause mold to grow or make inks bleed. In these cases, freeze-drying may not be an effective restoration method.

In bacteriology freeze-drying is used to conserve special strains.

In high-altitude environments, the low temperatures and pressures can sometimes produce natural mummies by a process of freeze-drying.

Advanced ceramics processes sometimes use freeze-drying to create a formable powder from a sprayed slurry mist. Freeze-drying creates softer particles with a more homogeneous chemical composition than traditional hot spray drying, but it is also more expensive.

Freeze-drying is also used for floral preservation. Wedding bouquet preservation has become very popular with brides who want to preserve their wedding day flowers

A new form of burial which previously freeze-dries the body with liquid nitrogen has been developed by the Swedish company Promessa Organic AB, which puts it forward as an environmentally friendly alternative to traditional casket and cremation burials.

Freeze-Drying Equipment

There are essentially three categories of freeze-dryers: the manifold freeze-dryer, the rotary freeze-dryer and the tray style freeze-dryer. Two components are common to all types of freeze-dryers: a vacuum pump to reduce the ambient gas pressure in a vessel containing the substance to be dried and a condenser to remove the moisture by condensation on a surface cooled to −40 to −80 °C (−40 to −112 °F). The manifold, rotary and tray type freeze-dryers differ in the method by which the dried substance is interfaced with a condenser. In manifold freeze-dryers a short usually circular tube is used to connect multiple containers with the dried product to a condenser. The rotary and tray freeze-dryers have a single large reservoir for the dried substance.

Unloading trays of freeze-dried material from a small cabinet-type freeze-dryer

Rotary freeze-dryers are usually used for drying pellets, cubes and other pourable substances. The rotary dryers have a cylindrical reservoir that is rotated during drying to achieve a more uniform drying throughout the substance. Tray style freeze-dryers usually have rectangular reservoir with shelves on which products, such as pharmaceutical solutions and tissue extracts, can be placed in trays, vials and other containers.

Manifold freeze-dryers are usually used in a laboratory setting when drying liquid substances in small containers and when the product will be used in a short period of time. A manifold dryer will dry the product to less than 5% moisture content. Without heat, only primary drying (removal of the unbound water) can be achieved. A heater must be added for secondary drying, which will remove the bound water and will produce a lower moisture content.

Tray style freeze-dryers are typically larger than the manifold dryers and are more sophisticated. Tray style freeze-dryers are used to dry a variety of materials. A tray freeze-dryer is used to produce the driest product for long-term storage. A tray freeze-dryer allows the product to be frozen in place and performs both primary (unbound water removal) and secondary (bound water removal) freeze-drying, thus producing the driest possible end-product. Tray freeze-dryers can dry products in bulk or in vials or other containers. When drying in vials, the freeze-dryer is supplied with a stoppering mechanism that allows a stopper to be pressed into place, sealing the vial before it is exposed to the atmosphere. This is used for long-term storage, such as vaccines.

Improved freeze-drying techniques are being developed to extend the range of products that can be freeze-dried, to improve the quality of the product, and to produce the product faster with less labor.

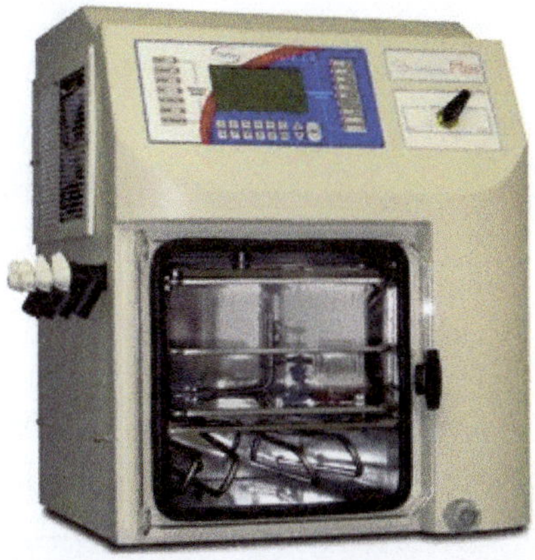

Product viewable single shelf freeze-dryer

Production freeze-drier

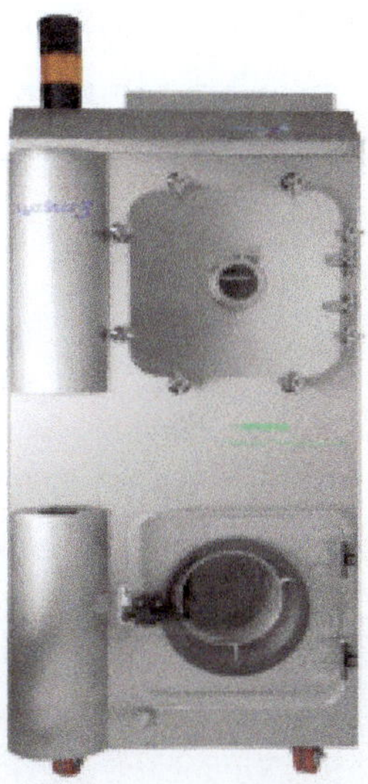

Development freeze-dryer

Future Food Technology

New food technologies can offer solutions to malnutrition. According to the WHO (World Healthcare Organization)approximately 30% of global population is malnourished. It is suggested that by 2020 the whole European Union will consume less food than China and India together.

Food Patch Plasters

These types of plasters are created with the cooperation of american scientists and militaries. C. Patrick Dann who works on this project promises that these plasters will appear on the market by 2025. At this stage of technological development they definitely cannot totally replace food consumption. However, at some extreme cases, such as battlefield or catastrophe, it can help people to survive for a longer time without proper food or water consumption. C. Patrick Dann also states that these kinds of plasters can be helpful for civilians e.g. miners.

Respirable Food

Harvard University professor David Edwards has invented a device called 'Le Whiff'. It helps to spray chocolate. Later this device was developed and on american market it appeared as 'Le Whaf'. Ten minutes of usage of such device provides 200 calories.

Printed Food

In 2013 NASA declared tht they are developing the type of food that can be printed. The main goal of the agency was to ensure that austronauts could print out food, instead of consuming it out of tubes. The conference SXSW demonstrated the first printed pizza.

Proteins Created from Insects

Morgaine Gaye a futurologist with specialization on food is confident that insects can easily replace cattle. She points out to the fact that around two billion of people regularly consume around two thousand types of insects and do well. Insects are rich with proteins and minerals, contain less fat than usual meat and reproduce quickly. Raise cattle is more difficult than insects. Moreover, raising insects does not harm environment as much as it harmed by cattle. It is also noted that fly larvae has a big potential.

References

- Potter, Norman N.; Hotchkiss, Joseph H. (1998). Food Science. Food science texts series (5th ed.). Springer. ISBN 9780834212657.
- Fratamico PM and Bayles DO (editor). (2005). Foodborne Pathogens: Microbiology and Molecular Biology. Caister Academic Press. ISBN 978-1-904455-00-4.
- Ljungh A, Wadstrom T (editors) (2009). Lactobacillus Molecular Biology: From Genomics to Probiotics. Caister Academic Press. ISBN 978-1-904455-41-7.
- Mayo, B; van Sinderen, D (editor) (2010). Bifidobacteria: Genomics and Molecular Aspects. Caister Academic Press. ISBN 978-1-904455-68-4.

- Hans-Jürgen Bässler und Frank Lehmann : Containment Technology: Progress in the Pharmaceutical and Food Processing Industry. Springer, Berlin 2013, ISBN 978-3642392917
- p 82 in "Lab Ref, Volume 2: A Handbook of Recipes, Reagents, and Other Reference Tools for Use at the Bench" by A.S. Mellick and L. Rodgers, CSHL Press, 2002, ISBN 978-0-87969-630-6
- В Рязани производят кошерное сухое молоко [Kosher Milk is Produced in Ryazan]. Наша Рязань (in Russian). 16 February 2010. Retrieved 22 April 2013.

3

Important Aspects of Food Science and Engineering

Food is grown, transported for processing, processed and finally packaged after which it is marketed to the consumer. An aspect that is crucial for the food industry is safe handling and storage of food. Food science and engineering comprises multiple aspects such as food processing, food preservation, food packaging, food marketing and storage. This chapter studies these aspects, the crucial roles they play and the techniques used in each.

Food Processing

Industrial cheese production

Food processing is the transformation of raw ingredients, by physical or chemical means into food, or of food into other forms. Food processing combines raw food ingredients to produce marketable food products that can be easily prepared and served by the consumer. Food processing typically involves activities such as mincing and macerating, liquefaction, emulsification, and cooking (such as boiling, broiling, frying, or grilling); pickling, pasteurization, and many other kinds of preservation; and canning or other packaging. (Primary-processing such as dicing or slicing, freezing or drying when leading to secondary products are also included.)

History

Food processing dates back to the prehistoric ages when crude processing incorporated fermenting, sun drying, preserving with salt, and various types of cooking (such as roasting, smoking, steaming, and oven baking), Such basic food processing involved chemical enzymatic changes to the basic structure of food in its natural form, as well served to build a barrier against surface microbial activity that caused rapid decay. Salt-preservation was especially common for foods that constituted warrior and sailors' diets until the introduction of canning methods. Evidence for the existence of these methods can be found in the writings of the ancient Greek, Chaldean, Egyptian and Roman civilizations as well as archaeological evidence from Europe, North and South America and Asia. These tried and tested processing techniques remained essentially the same until the advent of the industrial revolution. Examples of ready-meals also date back to before the preindustrial revolution, and include dishes such as Cornish pasty and Haggis. Both during ancient times and today in modern society these are considered processed foods.

Grain silos in Ardrossan, Scotland

Modern food processing technology developed in the 19th and 20th centuries was developed in a large part to serve military needs. In 1809 Nicolas Appert invented a hermetic bottling technique that would preserve food for French troops which ultimately contributed to the development of tinning, and subsequently canning by Peter Durand in 1810. Although initially expensive and somewhat hazardous due to the lead used in cans, canned goods would later become a staple around the world. Pasteurization, discovered by Louis Pasteur in 1864, improved the quality of preserved foods and introduced the wine, beer, and milk preservation.

A form of pre-made split-pea soup that has become traditional

In the 20th century, World War II, the space race and the rising consumer society in developed countries (including the United States) contributed to the growth of food processing with such advances as spray drying, juice concentrates, freeze drying and the introduction of artificial sweeteners, colouring agents, and such preservatives as sodium benzoate. In the late 20th century, products such as dried instant soups, reconstituted fruits and juices, and self cooking meals such as MRE food ration were developed.

In western Europe and North America, the second half of the 20th century witnessed a rise in the pursuit of convenience. Food processing companies marketed their products especially towards middle-class working wives and mothers. Frozen foods (often credited to Clarence Birdseye) found their success in sales of juice concentrates and "TV dinners". Processors utilised the perceived value of time to appeal to the postwar population, and this same appeal contributes to the success of convenience foods today.

Benefits and Drawbacks

Benefits

Processed seafood - fish, squid, prawn balls and simulated crab sticks (surimi)

Benefits of food processing include toxin removal, preservation, easing marketing and distribution tasks, and increasing food consistency. In addition, it increases yearly availability of many foods, enables transportation of delicate perishable foods across long distances and makes many kinds of foods safe to eat by de-activating spoilage and pathogenic micro-organisms. Modern supermarkets would not exist without modern food processing techniques, and long voyages would not be possible.

Processed foods are usually less susceptible to early spoilage than fresh foods and are better suited for long distance transportation from the source to the consumer. When they were first introduced, some processed foods helped to alleviate food shortages and improved the overall nutrition of populations as it made many new foods available to the masses.

Processing can also reduce the incidence of food borne disease. Fresh materials, such as fresh produce and raw meats, are more likely to harbour pathogenic micro-organisms (e.g. Salmonella) capable of causing serious illnesses.

The extremely varied modern diet is only truly possible on a wide scale because of food processing. Transportation of more exotic foods, as well as the elimination of much hard labour gives the modern eater easy access to a wide variety of food unimaginable to their ancestors.

The act of processing can often improve the taste of food significantly.

Mass production of food is much cheaper overall than individual production of meals from raw ingredients. Therefore, a large profit potential exists for the manufacturers and suppliers of processed food products. Individuals may see a benefit in convenience, but rarely see any direct financial cost benefit in using processed food as compared to home preparation.

Libby's brand 'Potted Meat Food Product'

Processed food freed people from the large amount of time involved in preparing and cooking "natural" unprocessed foods. The increase in free time allows people much more choice in life style than previously allowed. In many families the adults are working away from home and therefore there is little time for the preparation of food based on fresh ingredients. The food industry offers products that fulfill many different needs: e.g. fully prepared ready meals that can be heated up in the microwave oven within a few minutes.

Modern food processing also improves the quality of life for people with allergies, diabetics, and other people who cannot consume some common food elements. Food processing can also add extra nutrients such as vitamins.

Drawbacks

Meat packages in a Roman supermarket

Any processing of food can affect its nutritional density. The amount of nutrients lost depends on the food and processing method. For example, heat destroys vitamin C. Therefore, canned fruits possess less vitamin C than their fresh alternatives. The USDA conducted a study in 2004, creating a nutrient retention table for several foods.

New research highlighting the importance to human health of a rich microbial environment in the intestine indicates that abundant food processing (not fermentation of foods) endangers that environment.

Using food additives represents another safety concern. The health risks of any given additive vary greatly from person to person; for example using sugar as an additive endangers diabetics. In the European Union, only European Food Safety Authority (EFSA) approved food additives (e.g., sweeteners, preservatives, stabilizers) are permitted at specified levels for use in food products. Approved additives receive an E number (E for Europe), simplifying communication about food additives included in the ingredients' list for all the different languages spoken in the EU. Certain additives can also result in an addiction to a particular food item. As effects of chemical additives are learnt, changes to laws and regulatory practices are made to make such processed foods more safe.

Food processing is typically a mechanical process that utilizes large mixing, grinding, chopping and emulsifying equipment in the production process. These processes inherently introduce a number of contamination risks. As a mixing bowl or grinder is used over time the food contact parts will tend to fail and fracture. This type of failure will introduce into the product stream small to large metal contaminants. Further processing of these metal fragments will result in downstream equipment failure and the risk of ingestion by the consumer. Food manufacturers utilize industrial metal detectors to detect and reject automatically any metal fragment. Large food processors will utilize many metal detectors within the processing stream to reduce both damage to processing machinery as well as risk to consumer health.

Typical Maximum Nutrient Losses due to cooking					
Vitamin & Minerals	Freeze	Dry	Cook	Cook+Drain	Reheat
Calcium	5%	0%	20%	25%	0%
Copper	10%	0%	40%	45%	0%
Folate	5%	50%	70%	75%	30%
Folic Acid	5%	50%	70%	75%	30%
Food Folate	5%	50%	70%	75%	30%
Iron	0%	0%	35%	40%	0%
Magnesium	0%	0%	25%	40%	0%
Niacin	0%	10%	40%	55%	5%
Phosphorus	0%	0%	25%	35%	0%
Potassium	10%	0%	30%	70%	0%
Riboflavin	0%	10%	25%	45%	5%
Sodium	0%	0%	25%	55%	0%
Thiamin	5%	30%	55%	70%	40%
Vit A- Alpha Carotene	5%	50%	25%	35%	10%
Vit A- Beta Carotene	5%	50%	25%	35%	10%
Vit A- Beta Cryptoxanthin	5%	50%	25%	35%	10%
Vit A- Lutein+Zeaxanthin	5%	50%	25%	35%	10%
Vit A- Lycopene	5%	50%	25%	35%	10%
Vit A- Retinol Activity Equivalent	5%	50%	25%	35%	10%
Vitamin A	5%	50%	25%	35%	10%
Vitamin B12	0%	0%	45%	50%	45%

Typical Maximum Nutrient Losses due to cooking					
Vitamin & Minerals	Freeze	Dry	Cook	Cook+Drain	Reheat
Vitamin B6	0%	10%	50%	65%	45%
Vitamin C	30%	80%	50%	75%	50%
Zinc	0%	0%	25%	25%	0%

Performance Parameters for Food Processing

Factory Automation - Robotics Palettizing Bread

When designing processes for the food industry the following performance parameters may be taken into account:

- Hygiene, e.g. measured by number of micro-organisms per mL of finished product
- Energy efficiency measured e.g. by "ton of steam per ton of sugar produced"
- Minimization of waste, measured e.g. by "percentage of peeling loss during the peeling of potatoes"
- Labour used, measured e.g. by "number of working hours per ton of finished product"
- Minimization of cleaning stops measured e.g. by "number of hours between cleaning stops"

De-Agglomerating Batter Mixes in Food Processing

Problems often occur during preparation of batter mixes because flour and other powdered ingredients tend to form lumps or agglomerates as they are being mixed during production. A conventional mixer/agitator cannot break down these agglomerates, resulting in a lumpy batter. If lumpy batter is used to enrobe products, it causes an unsatisfactory appearance with misshapen or oversize products that do not fit properly into packaging. This can force production to a standstill. Furthermore, batter mix is generally recirculated from an enrobing system back to a holding vessel; lumps then have a tendency to build up, reducing the flow of material and raising potential sanitation issues.

Using a high shear in-line mixer in place of a conventional agitator or mixer can quickly solve problems of agglomeration with dry ingredients. A single pass through a self-pumping, in-line mixer adds high shear to batter, which de-agglomerates the mix, resulting in a homogeneous, smooth batter. With a consistent, smooth batter, finished product appearance is improved; the effectiveness and hygiene of the recirculation system is increased; and a better yield of raw materials is achieved. By increasing overall product quality, the amount of raw materials needed is decreased, thereby lowering manufacturing costs and increasing shelf life. Increased shelf life is achieved by creating and maintaining an emulsion, often by adding a food stabilizer.

High shear in-line mixers process food to be made faster and cheaper while increasing consistency of the finished food. Powder and liquid mixing systems are capable of rapidly incorporating large quantities of powders at high concentrations – agglomerate free and fully hydrated. Advances in technology have made processing equipment easy to clean, leading to a much safer processed food.

Trends in Modern Food Processing

Women working in a cannery

Dried bananas packaged in Ban Bang Krathum, Bang Krathum, Phitsanulok, Thailand

Health

Reduction of fat content in final product by using baking instead of deep-frying in the production of potato chips, another processed food.

Maintaining the natural taste of the product by using less artificial sweetener than was used before.

Hygiene

The rigorous application of industry and government endorsed standards to minimise possible risk and hazards. The international standard adopted is HACCP.

Efficiency

Rising energy costs lead to increasing usage of energy-saving technologies, e.g. frequency converters on electrical drives, heat insulation of factory buildings and heated vessels, energy recovery systems, keeping a single fish frozen all the way from China to Switzerland.

Factory automation systems (often Distributed control systems) reduce personnel costs and may lead to more stable production results.

Food Preservation

Food preservation involves preventing the growth of bacteria, fungi (such as yeasts), or other micro-organisms (although some methods work by introducing benign bacteria or fungi to the food), as well as slowing the oxidation of fats that cause rancidity. Food preservation may also include processes that inhibit visual deterioration, such as the enzymatic browning reaction in apples after they are cut during food preparation.

Many processes designed to preserve food will involve a number of food preservation methods. Preserving fruit by turning it into jam, for example, involves boiling (to reduce the fruit's moisture content and to kill bacteria, etc.), sugaring (to prevent their re-growth) and sealing within an airtight jar (to prevent recontamination). Some traditional methods of preserving food have been shown to have a lower energy input and carbon footprint, when compared to modern methods. However, some methods of food preservation are known to create carcinogens, and in 2015, the International Agency for Research on Cancer of the World Health Organization classified processed meat, i.e. meat that has undergone salting, curing, fermenting, and smoking, as "carcinogenic to humans".

Maintaining or creating nutritional value, texture and flavor is an important aspect of food preservation, although, historically, some methods drastically altered the character of the food being preserved. In many cases these changes have come to be seen as desirable qualities – cheese, yogurt and pickled onions being common examples.

Traditional Techniques

New techniques of food preservation became available to the home chef from the dawn of agriculture until the Industrial Revolution.

Drying

Drying is one of the oldest techniques used to hamper the decomposition of food products. As early as 12,000 B.C., Middle Eastern and Oriental cultures were drying foods using the power of the sun. Vegetables and fruit are naturally dried by the sun and wind, but "still houses" were built in areas that did not have enough sunlight to dry things. A fire would be built inside the building to provide the heat to dry the various fruits, vegetables, and herbs.

Cooling

Cooling preserves foods by slowing down the growth and reproduction of micro-organisms and the action of enzymes that cause food to rot. The introduction of commercial and domestic refrigerators drastically improved the diets of many in the Western world by allowing foods such as fresh fruit, salads and dairy products to be stored safely for longer periods, particularly during warm weather.

Freezing

Freezing is also one of the most commonly used processes, both commercially and domestically, for preserving a very wide range of foods, including prepared foods that would not have required freezing in their unprepared state. For example, potato waffles are stored in the freezer, but potatoes themselves require only a cool dark place to ensure many months' storage. Cold stores provide large-volume, long-term storage for strategic food stocks held in case of national emergency in many countries.

Boiling

Boiling liquid food items can kill any existing microbes. Milk and water are often boiled to kill any harmful microbes that may be present in them.

Heating

Heating to temperatures which are sufficient to kill microorganisms inside the food is a method used with perpetual stews. Milk is also boiled before storing to kill many microorganisms.

Salting

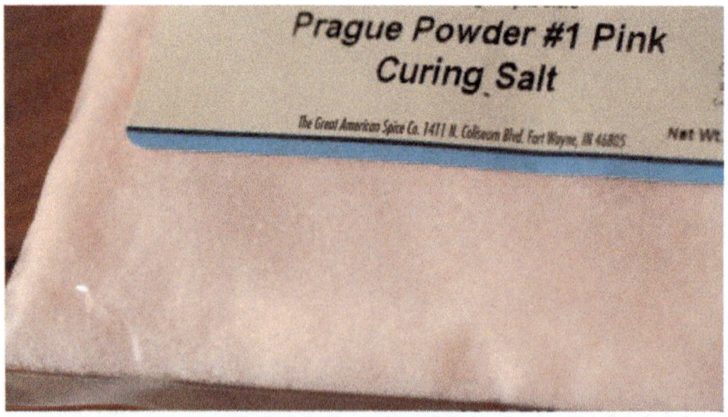

Bag of Prague powder #1, also known as "curing salt" or "pink salt". It is typically a combination of salt and sodium nitrite, with the pink color added to distinguish it from ordinary salt.

Salting or curing draws moisture from the meat through a process of osmosis. Meat is cured with salt or sugar, or a combination of the two. Nitrates and nitrites are also often used to cure meat and contribute the characteristic pink color, as well as inhibition of *Clostridium botulinum*. It was a main method of preservation in medieval times and around the 1700s.

Sugaring

The earliest cultures have used sugar as a preservative, and it was commonplace to store fruit in honey. Similar to pickled foods, sugar cane was brought to Europe through the trade routes. In northern climates without sufficient sun to dry foods, preserves are made by heating the fruit with sugar. "Sugar tends to draw water from the microbes (plasmolysis). This process leaves the microbial cells dehydrated, thus killing them. In this way, the food will remain safe from microbial spoilage." Sugar is used to preserve fruits, either in an anti-microbial syrup with fruit such as apples, pears, peaches, apricots and plums, or in crystallized form where the preserved material is cooked in sugar to the point of crystallization and the resultant product is then stored dry. This method is used for the skins of citrus fruit (candied peel), angelica and ginger. Also sugaring can be used in jam jellies.

Smoking

Smoked ham in a smokehouse in Schleswig-Holstein, Germany

Smoking is used to lengthen the shelf life of perishable food items. This effect is achieved by exposing the food to smoke from burning plant materials such as wood. Smoke deposits a number of pyrolysis products onto the food, including the phenols syringol, guaiacol and catechol. These compounds aid in the drying and preservation of meats and other foods. Most commonly subjected to this method of food preservation are meats and fish that have undergone curing. Fruits and vegetables like paprika, cheeses, spices, and ingredients for making drinks such as malt and tea leaves are also smoked, but mainly for cooking or flavoring them. It is one of the oldest food preservation methods, which probably arose after the development of cooking with fire.

Pickling

Pickling is a method of preserving food in an edible anti-microbial liquid. Pickling can be broadly classified into two categories: chemical pickling and fermentation pickling.

In chemical pickling, the food is placed in an edible liquid that inhibits or kills bacteria and other micro-organisms. Typical pickling agents include brine (high in salt), vinegar, alcohol, and vegetable oil, especially olive oil but also many other oils. Many chemical pickling processes also involve heating or boiling so that the food being preserved becomes saturated with the pickling agent. Common chemically pickled foods include cucumbers, peppers, corned beef, herring, and eggs, as well as mixed vegetables such as piccalilli.

In fermentation pickling, the food itself produces the preservation agent, typically by a process that produces lactic acid. Fermented pickles include sauerkraut, nukazuke, kimchi, surströmming, and curtido. Some pickled cucumbers are also fermented.

Lye

Sodium hydroxide (lye) makes food too alkaline for bacterial growth. Lye will saponify fats in the food, which will change its flavor and texture. Lutefisk uses lye in its preparation, as do some olive recipes. Modern recipes for century eggs also call for lye.

Canning

Canning involves cooking food, sealing it in sterile cans or jars, and boiling the containers to kill or weaken any remaining bacteria as a form of sterilization. It was invented by the French confectioner Nicolas Appert. By 1806, this process was used by the French Navy to preserve meat, fruit, vegetables, and even milk. Although Appert had discovered a new way of preservation, it wasn't understood until 1864 when Louis Pasteur found the relationship between microorganisms, food spoilage, and illness.

Preserved food

Foods have varying degrees of natural protection against spoilage and may require that the final step occur in a pressure cooker. High-acid fruits like strawberries require no preservatives to can and only a short boiling cycle, whereas marginal vegetables such as carrots require longer boiling and addition of other acidic elements. Low-acid foods, such as vegetables and meats, require pressure canning. Food preserved by canning or bottling is at immediate risk of spoilage once the can or bottle has been opened.

Lack of quality control in the canning process may allow ingress of water or micro-organisms. Most such failures are rapidly detected as decomposition within the can causes gas production and the can will swell or burst. However, there have been examples of poor manufacture (underprocessing) and poor hygiene allowing contamination of canned food by the obligate anaerobe *Clostridium botulinum*, which produces an acute toxin within the food, leading to severe illness or death. This organism produces no gas or obvious taste and remains undetected by taste or smell. Its toxin is denatured by cooking, however. Cooked mushrooms, handled poorly and then canned, can support the growth of Staphylococcus aureus, which produces a toxin that is not destroyed by canning or subsequent reheating.

Jellying

Food may be preserved by cooking in a material that solidifies to form a gel. Such materials include gelatin, agar, maize flour, and arrowroot flour. Some foods naturally form a protein gel when cooked, such as eels and elvers, and sipunculid worms, which are a delicacy in Xiamen, in the Fujian province of the People's Republic of China. Jellied eels are a delicacy in the East End of London, where they are eaten with mashed potatoes. Potted meats in aspic (a gel made from gelatine and clarified meat broth) were a common way of serving meat off-cuts in the UK until the 1950s. Many jugged meats are also jellied.

A traditional British way of preserving meat (particularly shrimp) is by setting it in a pot and sealing it with a layer of fat. Also common is potted chicken liver; compare pâté.

Jugging

Meat can be preserved by jugging. Jugging is the process of stewing the meat (commonly game or fish) in a covered earthenware jug or casserole. The animal to be jugged is usually cut into pieces, placed into a tightly-sealed jug with brine or gravy, and stewed. Red wine and/or the animal's own blood is sometimes added to the cooking liquid. Jugging was a popular method of preserving meat up until the middle of the 20th century.

Burial

Burial of food can preserve it due to a variety of factors: lack of light, lack of oxygen, cool temperatures, pH level, or desiccants in the soil. Burial may be combined with other methods such as salting or fermentation. Most foods can be preserved in soil that is very dry and salty (thus a desiccant) such as sand, or soil that is frozen.

Many root vegetables are very resistant to spoilage and require no other preservation than storage in cool dark conditions, for example by burial in the ground, such as in a storage clamp. Century eggs are created by placing eggs in alkaline mud (or other alkaline substance), resulting in their "inorganic" fermentation through raised pH instead of spoiling. The fermentation preserves them and breaks down some of the complex, less flavorful proteins and fats into simpler, more flavorful ones. Cabbage was traditionally buried in the fall in northern farms in the U.S. for preservation. Some methods keep it crispy while other methods produce sauerkraut. A similar process is used in the traditional production of kimchi. Sometimes meat is buried under conditions that cause preservation. If buried on hot coals or ashes, the heat can kill pathogens, the dry ash can desiccate, and the earth can block oxygen and further contamination. If buried where the earth is very cold, the earth acts like a refrigerator.

In Orissa, India, it is practical to store rice by burying it underground. This method helps to store for three to six months during the dry season.

Curing

The earliest form of curing was dehydration. To accelerate this process, salt is usually added. In the culinary world, it was common to choose raw salts from various sources (rock salt, sea salt, etc.).

More modern "examples of salts that are used as preservatives include sodium chloride (NaCl), sodium nitrate ($NaNO_3$) and sodium nitrite ($NaNO_2$). Even at mild concentrations (up to 2%), sodium chloride, found in many food products, is capable of neutralizing the antimicrobial character of natural compounds."

Fermentation

Some foods, such as many cheeses, wines, and beers, use specific micro-organisms that combat spoilage from other less-benign organisms. These micro-organisms keep pathogens in check by creating an environment toxic for themselves and other micro-organisms by producing acid or alcohol. Methods of fermentation include, but are not limited to, starter micro-organisms, salt, hops, controlled (usually cool) temperatures and controlled (usually low) levels of oxygen. These methods are used to create the specific controlled conditions that will support the desirable organisms that produce food fit for human consumption.

Fermentation is the microbial conversion of starch and sugars into alcohol. Not only can fermentation produce alcohol, but it can also be a valuable preservation technique. Fermentation can also make foods more nutritious and palatable. For example, drinking water in the Middle Ages was dangerous because it often contained pathogens that could spread disease. When the water is made into beer, the boiling during the brewing process kills any bacteria in the water that could make people sick. Additionally, the water now has the nutrients from the barley and other ingredients, and the microorganisms can also produce vitamins as they ferment.

Industrial/Modern Techniques

Techniques of food preservation were developed in research laboratories for commercial applications.

Pasteurization

Pasteurization is a process for preservation of liquid food. It was originally applied to combat the souring of young local wines. Today, the process is mainly applied to dairy products. In this method, milk is heated at about 70 °C for 15 to 30 seconds to kill the bacteria present in it and cooling it quickly to 10 °C to prevent the remaining bacteria from growing. The milk is then stored in sterilized bottles or pouches in cold places. This method was invented by Louis Pasteur, a French chemist, in 1862.

Vacuum Packing

Vacuum-packing stores food in a vacuum environment, usually in an air-tight bag or bottle. The vacuum environment strips bacteria of oxygen needed for survival. Vacuum-packing is commonly used for storing nuts to reduce loss of flavor from oxidization. A major drawback to vacuum packaging, at the consumer level, is that vacuum sealing can deform contents and rob certain foods, such as cheese, of its flavor.

Artificial Food Additives

Preservative food additives can be *antimicrobial*, which inhibit the growth of bacteria or fungi, including mold, or *antioxidant*, such as oxygen absorbers, which inhibit the oxidation of food constituents. Common antimicrobial preservatives include calcium propionate, sodium nitrate, sodium nitrite, sulfites (sulfur dioxide, sodium bisulfite, potassium hydrogen sulfite, etc.) and disodium EDTA. Antioxidants include BHA and BHT. Other preservatives include formaldehyde (usually in solution), glutaraldehyde (kills insects), ethanol, and methylchloroisothiazolinone.

Irradiation

Irradiation of food is the exposure of food to ionizing radiation. The two types of ionizing radiation used are beta particles (high-energy electrons) and gamma rays (emitted from radioactive sources as cobalt-60 or cesium-137). Treatment effects include killing bacteria, molds, and insect pests, reducing the ripening and spoiling of fruits, and at higher doses inducing sterility. The technology may be compared to pasteurization; it is sometimes called "cold pasteurization", as the product is not heated.

The irradiation process is not directly related to nuclear energy, but does use radioactive isotopes produced in nuclear reactors. Cobalt-60, for example does not occur naturally and can only be produced through neutron bombardment of cobalt-59. Ionizing radiation at high energy levels is hazardous to life (hence its usefulness in sterilisation); for this reason, irradiation facilities have a heavily shielded irradiation room where the process takes place. Radiation safety procedures are used to ensure that neither the workers in such facilities nor the environment receives any radiation dose above administrative limits. Irradiated food does not and cannot become radioactive, and national and international expert bodies have declared food irradiation as wholesome. However, the wholesomeness of consuming such food is disputed by opponents and consumer organizations. National and international expert bodies have declared food irradiation as "wholesome"; organizations of the United Nations, such as the World Health Organization and Food and Agriculture Organization, endorse food irradiation. International legislation on whether food may be irradiated or not varies worldwide from no regulation to full banning. Irradiation may allow lower-quality or contaminated foods to be rendered marketable.

Approximately 500,000 tons of food items are irradiated per year worldwide in over 40 countries. These are mainly spices and condiments with an increasing segment of fresh fruit irradiated for fruit fly quarantine.

Pulsed Electric Field Electroporation

Pulsed electric field (PEF) electroporation is a method for processing cells by means of brief pulses of a strong electric field. PEF holds potential as a type of low-temperature alternative pasteurization process for sterilizing food products. In PEF processing, a substance is placed between two electrodes, then the pulsed electric field is applied. The electric field enlarges the pores of the cell membranes, which kills the cells and releases their contents. PEF for food processing is a developing technology still being researched. There have been limited industrial applications of PEF processing for the pasteurization of fruit juices. To date, several PEF treated juices are available on the market in Europe. Furthermore, for several years a juice pasteurization application in the US has used PEF. For cell disintegration purposes especially potato processors show great interest in PEF technology as an efficient alternative for their preheaters. Potato applications are already operational in the US and Canada. There are also commercial PEF potato applications in various countries in Europe, as well as in Australia, India and China.

Modified Atmosphere

Modifying atmosphere is a way to preserve food by operating on the atmosphere around it. Salad crops that are notoriously difficult to preserve are now being packaged in sealed bags with an atmosphere modified to reduce the oxygen (O_2) concentration and increase the carbon dioxide (CO_2) concentration. There is concern that, although salad vegetables retain their appearance and texture in such conditions, this method of preservation may not retain nutrients, especially vitamins. There are two methods for preserving grains with carbon dioxide. One method is placing a block of dry ice in the bottom and filling the can with the grain. Another method is purging the container from the bottom by gaseous carbon dioxide from a cylinder or bulk supply vessel.

Carbon dioxide prevents insects and, depending on concentration, mold and oxidation from damaging the grain. Grain stored in this way can remain edible for approximately five years.

Nitrogen gas (N_2) at concentrations of 98% or higher is also used effectively to kill insects in the grain through hypoxia. However, carbon dioxide has an advantage in this respect, as it kills organisms through hypercarbia and hypoxia (depending on concentration), but it requires concentrations of above 35%, or so. This makes carbon dioxide preferable for fumigation in situations where a hermetic seal cannot be maintained.

Controlled Atmospheric Storage (CA): "CA storage is a non-chemical process. Oxygen levels in the sealed rooms are reduced, usually by the infusion of nitrogen gas, from the approximate 21 percent in the air we breathe to 1 percent or 2 percent. Temperatures are kept at a constant 0 to 2 °C (32 to 36 °F). Humidity is maintained at 95 percent and carbon dioxide levels are also controlled. Exact conditions in the rooms are set according to the apple variety. Researchers develop specific regimens for each variety to achieve the best quality. Computers help keep conditions constant." "Eastern Washington, where most of Washington's apples are grown, has enough warehouse storage for 181 million boxes of fruit, according to a report done in 1997 by managers for the Washington State Department of Agriculture Plant Services Division. The storage capacity study shows that 67 percent of that space —enough for 121,008,000 boxes of apples — is CA storage."

Air-tight storage of grains (sometimes called hermetic storage) relies on the respiration of grain, insects, and fungi that can modify the enclosed atmosphere sufficiently to control insect pests. This is a method of great antiquity, as well as having modern equivalents. The success of the method relies on having the correct mix of sealing, grain moisture, and temperature.

A patented process uses fuel cells to exhaust and automatically maintain the exhaustion of oxygen in a shipping container, containing, for example, fresh fish.

Nonthermal Plasma

This process subjects the surface of food to a "flame" of ionized gas molecules, such as helium or nitrogen. This causes micro-organisms to die off on the surface.

High-Pressure Food Preservation

High-pressure food preservation or pascalization refers to the use of a food preservation technique that makes use of high pressure. "Pressed inside a vessel exerting 70,000 pounds per square inch

(480 MPa) or more, food can be processed so that it retains its fresh appearance, flavor, texture and nutrients while disabling harmful microorganisms and slowing spoilage." By 2005, the process was being used for products ranging from orange juice to guacamole to deli meats and widely sold.

Biopreservation

Biopreservation is the use of natural or controlled microbiota or antimicrobials as a way of preserving food and extending its shelf life. Beneficial bacteria or the fermentation products produced by these bacteria are used in biopreservation to control spoilage and render pathogens inactive in food. It is a benign ecological approach which is gaining increasing attention.

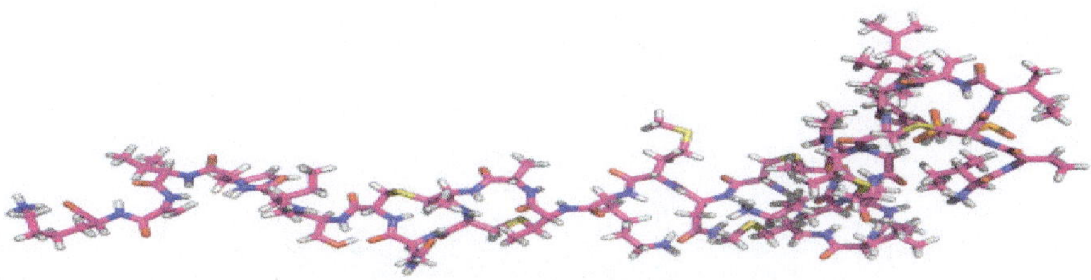

3D stick model of nisin. Some lactic acid bacteria manufacture nisin. It is a particularly effective preservative.

Of special interest are lactic acid bacteria (LAB). Lactic acid bacteria have antagonistic properties that make them particularly useful as biopreservatives. When LABs compete for nutrients, their metabolites often include active antimicrobials such as lactic acid, acetic acid, hydrogen peroxide, and peptide bacteriocins. Some LABs produce the antimicrobial nisin, which is a particularly effective preservative.

These days, LAB bacteriocins are used as an integral part of hurdle technology. Using them in combination with other preservative techniques can effectively control spoilage bacteria and other pathogens, and can inhibit the activities of a wide spectrum of organisms, including inherently resistant Gram-negative bacteria.

Hurdle Technology

Hurdle technology is a method of ensuring that pathogens in food products can be eliminated or controlled by combining more than one approach. These approaches can be thought of as "hurdles" the pathogen has to overcome if it is to remain active in the food. The right combination of hurdles can ensure all pathogens are eliminated or rendered harmless in the final product.

Hurdle technology has been defined by Leistner (2000) as an intelligent combination of hurdles that secures the microbial safety and stability as well as the organoleptic and nutritional quality and the economic viability of food products. The organoleptic quality of the food refers to its sensory properties, that is its look, taste, smell, and texture.

Examples of hurdles in a food system are high temperature during processing, low temperature during storage, increasing the acidity, lowering the water activity or redox potential, and the presence of preservatives or biopreservatives. According to the type of pathogens and how risky they are, the intensity of the hurdles can be adjusted individually to meet consumer preferences in an economical way, without sacrificing the safety of the product.

Principal hurdles used for food preservation (after Leistner, 1995)		
Parameter	Symbol	Application
High temperature	F	Heating
Low temperature	T	Chilling, freezing
Reduced water activity	a_w	Drying, curing, conserving
Increased acidity	pH	Acid addition or formation
Reduced redox potential	E_h	Removal of oxygen or addition of ascorbate
Biopreservatives		Competitive flora such as microbial fermentation
Other preservatives		Sorbates, sulfites, nitrites

Food Industry

Packaged food aisles at an American grocery store

Parmigiano reggiano Cheese in a modern factory

Hens in a battery cage in Brazil, an example of intensive animal farming

The food industry is a complex, global collective of diverse businesses that supply most of the food consumed by the world population. Only subsistence farmers, those who survive on what they grow, can be considered outside of the scope of the modern food industry.

The Food Industry Includes:

- Agriculture: raising of crops and livestock, and seafood
- Manufacturing: agrichemicals, agricultural construction, farm machinery and supplies, seed, etc.
- Food processing: preparation of fresh products for market, and manufacture of prepared food products
- Marketing: promotion of generic products (e.g., milk board), new products, advertising, marketing campaigns, packaging, public relations, etc.
- Wholesale and distribution: logistics, transportation, warehousing
- Foodservice (which includes Catering)
- Grocery, farmers' markets, public markets and other retailing
- Regulation: local, regional, national, and international rules and regulations for food production and sale, including food quality, food safety, marketing/advertising, and industry lobbying activities
- Education: academic, consultancy, vocational
- Research and development: food technology
- Financial services: credit, insurance

Definitions

It is challenging to find an inclusive way to cover all aspects of food production and sale. The Food Standards Agency, a government body in India, describes it thusly:

"...the whole food industry – from farming and food production, packaging and distribution, to retail and catering."

The Economic Research Service of the USDA uses the term *food system* to describe the same thing:

"The U.S. food system is a complex network of farmers and the industries that link to them. Those links include makers of farm equipment and chemicals as well as firms that provide services to agribusinesses, such as providers of transportation and financial services. The system also includes the food marketing industries that link farms to consumers, and which include food and fiber processors, wholesalers, retailers, and foodservice establishments."

Agriculture and Agronomy

Agriculture is the process of producing food, feeding products, fiber and other desired products by the cultivation of certain plants and the raising of domesticated animals (livestock). The practice of agriculture is also known as "farming". Scientists, inventors, and others devoted to improving farming methods and implements are also said to be engaged in agriculture. 1 in 3 people worldwide are employed in agriculture, yet it only contributes 3% to global GDP.

A soybean field in Junin, Argentina

Agronomy is the science and technology of producing and using plants for food, fuel, fibre, and land reclamation. Agronomy encompasses work in the areas of plant genetics, plant physiology, meteorology, and soil science. Agronomy is the application of a combination of sciences. Agronomists today are involved with many issues including producing food, creating healthier food, managing environmental impact of agriculture, and extracting energy from plants.

Food Processing

Packaged meat in a supermarket

Food processing includes the methods and techniques used to transform raw ingredients into food for human consumption. Food processing takes clean, harvested or slaughtered and butchered components and uses them to produce marketable food products. There are several different ways in which food can be produced.

One off production: This method is used when customers make an order for something to be made to their own specifications, for example a wedding cake. The making of one-off products could take days depending on how intricate the design is.

Batch production: This method is used when the size of the market for a product is not clear, and where there is a range within a product line. A certain number of the same goods will be produced to make up a batch or run, for example a bakery may bake a limited number of cupcakes. This method involves estimating consumer demand.

Mass production: This method is used when there is a mass market for a large number of identical products, for example chocolate bars, ready meals and canned food. The product passes from one stage of production to another along a production line.

Just-in-time (JIT) (production): This method of production is mainly used in restaurants. All components of the product are available in-house and the customer chooses what they want in the product. It is then prepared in a kitchen, or in front of the buyer as in sandwich delicatessens, pizzerias, and sushi bars.

Wholesale and Distribution

A vast global cargo network connects the numerous parts of the industry. These include suppliers, manufacturers, warehousers, retailers and the end consumers. Wholesale markets for fresh food products have tended to decline in importance in urbanizing countries, including Latin America and some Asian countries as a result of the growth of supermarkets, which procure directly from farmers or through preferred suppliers, rather than going through markets.

A foodservice truck at a loading dock. Trucks commonly distribute food products to commercial businesses and organizations.

The constant and uninterrupted flow of product from distribution centers to store locations is a critical link in food industry operations. Distribution centers run more efficiently, throughput can be increased, costs can be lowered, and manpower better utilized if the proper steps are taken when setting up a material handling system in a warehouse.

Retail

With worldwide urbanization, food buying is increasingly removed from food production. During the 20th century, the supermarket became the defining retail element of the food industry. There, tens of thousands of products are gathered in one location, in continuous, year-round supply.

Food preparation is another area where the change in recent decades has been dramatic. Today, two food industry sectors are in apparent competition for the retail food dollar. The grocery industry sells fresh and largely raw products for consumers to use as ingredients in home cooking. The food service industry by contrast offers prepared food, either as finished products, or as partially prepared components for final "assembly". Restaurants, cafes, bakeries and mobile food trucks provide opportunities for consumers to purchase food.

Food Industry Technologies

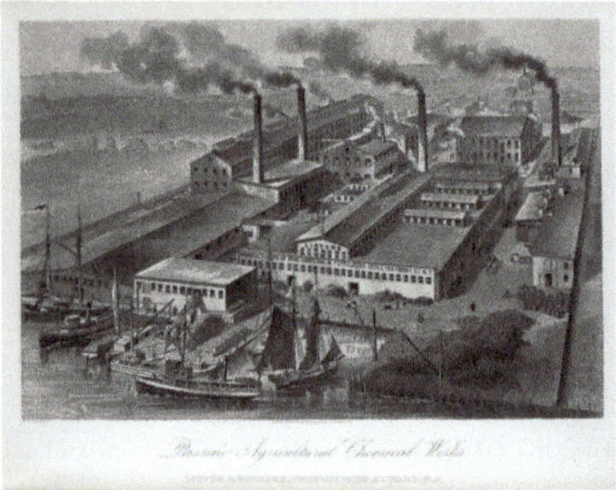

The Passaic Agricultural Chemical Works, an agrochemical company, in Newark, New Jersey, 1876

Modern food production is defined by sophisticated technologies. These include many areas. Agricultural machinery, originally led by the tractor, has practically eliminated human labor in many areas of production. Biotechnology is driving much change, in areas as diverse as agrochemicals, plant breeding and food processing. Many other types of technology are also involved, to the point where it is hard to find an area that does not have a direct impact on the food industry. As in other fields, computer technology is also a central force, with computer networks and specialized software providing the support infrastructure to allow global movement of the myriad components involved.

Marketing

As consumers grow increasingly removed from food production, the role of product creation, advertising, and publicity become the primary vehicles for information about food. With processed food as the dominant category, marketers have almost infinite possibilities in product creation.

Labor and Education

Some equipment at Tartu Mill, the largest grain milling company in the Baltic states. Modern food processing factories are often highly automated and need few workers.

Until the last 100 years, agriculture was labor intensive. Farming was a common occupation and millions of people were involved in food production. Farmers, largely trained from generation to generation, carried on the family business. That situation has changed dramatically today. In

America in 1870, 70-80 percent of the US population was employed in agriculture. As of 2008, less than 2 percent of the population is directly employed in agriculture., and about 80% of the population lives in cities. The food industry as a complex whole requires an incredibly wide range of skills. Several hundred occupation types exist within the food industry.

Food Packaging

Food packaging is packaging for food. A package provides protection, tampering resistance, and special physical, chemical, or biological needs. It may bear a nutrition facts label and other information about food being offered for sale.

Testing modified atmosphere in a plastic bag of carrots

Functions

Packaging and package labeling have several objectives

- Physical protection - The food enclosed in the package may require protection from, among other things, shock, vibration, compression, temperature, bacteria, etc.
- Barrier protection - A barrier from oxygen, water vapor, dust, etc., is often required. Permeation is a critical factor in design. Some packages contain desiccants or oxygen absorbers to help extend shelf life. Modified atmospheres or controlled atmospheres are also maintained in some food packages. Keeping the contents clean, fresh, and safe for the intended shelf life is a primary function.
- Containment or agglomeration - Small items are typically grouped together in one package to allow efficient handling. Liquids, powders, and granular materials need containment.
- Information transmission - Packages and labels communicate how to use, transport, recycle, or dispose of the package or product. Some types of information are required by governments.

- Marketing - The packaging and labels can be used by marketers to encourage potential buyers to purchase the product. Package design has been an important and constantly evolving phenomenon for several decades. Marketing communications and graphic design are applied to the surface of the package and (in many cases) the point of sale display.
- Security - Packaging can play an important role in reducing the security risks of shipment. Packages can be made with improved tamper resistance to deter tampering and also can have tamper-evident features to help indicate tampering. Packages can be engineered to help reduce the risks of package pilferage; some package constructions are more resistant to pilferage and some have pilfer-indicating seals. Packages may include authentication seals to help indicate that the package and contents are not counterfeit. Packages also can include anti-theft devices, such as dye packs, RFID tags, or electronic article surveillance tags, that can be activated or detected by devices at exit points and require specialized tools to deactivate. Using packaging in this way is a means of retail loss prevention.
- Convenience - Packages can have features which add convenience in distribution, handling, stacking, display, sale, opening, reclosing, use, and reuse.
- Portion control - Single-serving packaging has a precise amount of contents to control usage. Bulk commodities (such as salt) can be divided into packages that are a more suitable size for individual households. It also aids the control of inventory: selling sealed one-liter bottles of milk, rather than having people bring their own bottles to fill themselves.

Types

The above materials are fashioned into different types of food packages and containers such as:

Packaging Type		
Aseptic processing	Primary	Liquid whole eggs or dairy products
Trays	Primary	Portion of fish or meat
Bags	Primary	Potato chips, apples, rice
Boxes	Secondary	Corrugated box of primary packages: box of cereal cartons, frozen pizzas
Cans	Primary	Can of tomato soup
Cartons	Primary	Carton of eggs, milk or juice cartons
Flexible packaging	Primary	Bagged salad
Pallets	Tertiary	A series of boxes on a single pallet used to transport from the manufacturing plant to a distribution center
Wrappers	Tertiary	Used to wrap the boxes on the pallet for transport

Primary packaging is the main package that holds the food that is being processed. Secondary packaging combines the primary packages into one box being made. Tertiary packaging combines all of the secondary packages into one pallet.

Packaging Machines

Shrink-wrapping trays of bakery goods

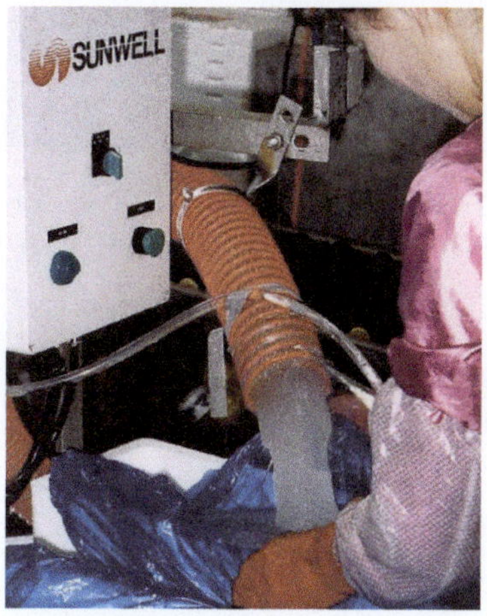

Pumping slurry ice onto fresh fish

A choice of packaging machinery requires consideration of technical capabilities, labor requirements, worker safety, maintainability, serviceability, reliability, ability to integrate into the packaging line, capital cost, floorspace, flexibility (change-over, materials, etc.), energy usage, quality of outgoing packages, qualifications (for food, pharmaceuticals, etc.), throughput, efficiency, productivity, and ergonomics, at a minimum.

- Packaging machines may be of the following general types:
- Autocoding label and date verification
- Blister-, skin- and vacuum-packaging machines
- Capping, over-capping, lidding, closing, seaming and sealing machines
- Cartoning machines
- Case and tray forming, packing, unpacking, closing and sealing Machines
- Check weighing machines

- Cleaning, sterilizing, cooling and drying machines
- Conveying and accumulating machines
- Feeding, orienting, and placing machines
- Filling machines for liquid and powdered products
- Package filling and closing Machines
- Form, fill and seal machines
- Inspecting, detecting and checkweighing machines
- Palletizing, depalletizing, and pallet unitizing machines
- Labeling, marking, and other product identification machines
- Wrapping machines
- Converting machines

Reducing Food Packaging

Reduced packaging and sustainable packaging are becoming more frequent. The motivations can be government regulations, consumer pressure, retailer pressure, and cost control. Reduced packaging often saves packaging costs.

In the UK, a Local Government Association survey produced by the British Market Research Bureau compared a range of outlets to buy 29 common food items and found that small local retailers and market traders "produced less packaging and more that could be recycled than the larger supermarkets."

Recycling of Food Packaging

After use, organic matter that is still in the food packaging needs to be separated from the packaging. This may also require rinsing of the food packaging.

Trends in Food Packaging

- Numerous reports industry associations agree that use of smart indicators will increase. There are a number of different indicators with different benefits for food producers, consumers and retailers.
- Temperature recorders are used to monitor products shipped in a cold chain and to help validate the cold chain. Digital temperature data loggers measure and record the temperature history of food shipments. They sometimes have temperatures displayed on the indicator or have other output (lights, etc.): The data from a shipment can be downloaded (cable, RFID, etc.) to a computer for further analysis. These help identify if there has been temperature abuse of products and can help determine the remaining shelf life. They can also help determine the time of temperature extremes during shipment so corrective measures can be taken.

    - Time temperature indicators integrate the time and temperature experienced by the indicator and adjacent foods. Some use chemical reactions that result in a color change while others use the migration of a dye through a filter media. To the degree that these physical changes in the indicator match the degradation rate of the food, the indicator can help indicate probable food degradation.

- Radio Frequency Identification is applied to food packages for supply chain control and has shown a significant benefit in allowing food producers and retailers create full real time visibility of their supply chain.
- Plastic packaging being used is usually non-biodegradable due to possible interactions with the food. Also, biodegradable polymers often require special composting conditions to properly degrade. Normal sealed landfill conditions do not promote biodegradation. Biodegradable plastics includes biodegradable films and coatings synthesized from organic materials and microbial polymers. Some package materials are edible. For example, pharmaceuticals are sometimes in capsules made of gelatin, starch, potato or other materials. Newer bioplastics, films and products are being developed.
- Barcodes have been used for decades in packaging many products. 2D barcodes used in Autocoding are increasingly applied to food packaging to ensure products are correctly packaged and date coded.
- The ability of a package to fully empty or dispense a viscous food is somewhat dependent on the surface energy of the inner walls of the container. The use of superhydrophobic surfaces is useful but can be further improved by using new lubricant-impregnated surfaces.

Food Safety and Public Health

It is critical to maintain food safety during processing , packaging, storage, logistics (including cold chain), sale, and use. Conformance to applicable regulations is mandatory. Some are country specific such as the US Food and Drug Administration and the US Department of Agriculture; others are regional such as the European Food Safety Authority. Certification programs such as the Global Food Safety Initiative are sometimes used. Food packaging considerations may include: use of hazard analysis and critical control points, verification and validation protocols, Good manufacturing practices, use of an effective quality management system, track and trace systems, and requirements for label content. Special food contact materials are used when the package is in direct contact with the food product. Depending on the packaging operation and the food, packaging machinery often needs specified daily wash-down and cleaning procedures.

Health risks of materials and chemicals used in food packaging need to be carefully controlled. Carcinogens, toxic chemicals, mutagens etc. need to be eliminated from food contact and potential migration into foods.

Food Marketing

Food marketing brings together the food producer and the consumer through a chain of marketing activities. The marketing of even a single food product can be a complicated process involvin g many producers and companies. For example, fifty-six companies are involved in making one can of chicken noodle soup. These businesses include not only chicken and vegetable processors but also the companies that transport the ingredients and those who print labels and manufacture cans. The food marketing system is the largest direct and indirect nongovernment employer in the United States.

Squashes and a pumpkin bought in the UK.

Food marketing is defined as a chain of marketing activities that take place within the food system between a food organisation and the consumer (Pomeranz & Adler, 2015). This has the potential to be a complicated procedure, as there are many processes that are used prior to the sale the food product. These include processing, wholesaling, retailing, food service and transport (Hawkes, 2007). Due to these many processes, a multitude of organisations have to be involved in the sale of one food product (Chandon, 2012). For example, approximately fifty-six organisations are involved in the making of one can of chicken noodle soup (Pomeranz & Adler, 2015). These organisations not only include the processors who make the ingredients for the product, but also involve the companies who manufacture the cans, print the labels and transport the product (Son, 2015). Therefore, on a global scale, the food marketing industry is one of the largest direct and indirect employers (Pomeranz & Adler, 2015).

Food marketing is the act of communicating to the consumer through a range of marketing techniques in order to add value to a food product and persuade the consumer to purchase. This includes all activities that occur in between the completion of a product through to the purchasing process of consumers. (Schaffner, Schroder & Earle 1997). Food marketing systems differ worldwide due to the level of development in the particular country, economically and technologically (Kaynak, 1999). Understanding and interpreting a particular countries food marketing techniques also requires taking into account the socio-economic, cultural, legal-political and technological environment of that country (Kaynak, 1999).

History

There are three historical phases of food marketing: the fragmentation phase (before 1870–1880), the unification phase (1880–1950), and the segmentation phase (1950 and later). In the fragmentation phase, the United States was divided into numerous geographic fragments because transporting food was expensive, leaving most production, distribution, and selling locally based. In the unification phase, distribution was made possible by railroads, coordination of sales forces was made possible by the telegraph and telephone, and product consistency was made possible by advances in manufacturing. This new distribution system was led by meat processors such as Armour and Swift in midwestern cities and by companies such as Heinz, Quaker Oats, Campbell Soup, and Coca-Cola, which sold their brands nationally. Advertising in print media and direct marketing through demonstrations at stores and public venues were among the prime marketing tools. The initial Crisco campaign, in 1911, was an example. In the segmentation phase (1950 and later) radio, television and internet advertising made it possible for a wider range of competing products to focus on different benefits and images and thus appeal to different demographic and psychographic markets. Distribution via the new national road system strengthened national brands.

History in The United States (U.S.)

The fragmentation phase (pre-1870 - 1880)

At the time of this phase, the production, distribution and sale of food products was based locally, as the transportation of food in the U.S. was expensive (Murata, 2007). This was achieved by dividing the country into numerous segments, making it cheaper for food producers (Son, 2015).

The unification phase (1880 - 1950)

During this phase, the use of railroads made food distribution possible; telephones aided the co-ordination of sales; and advances in manufacturing technology made product consistency possible (Son, 2015). The prime marketing tools for this phase were advertising aspects, including print media, and direct marketing through store demonstrations (Murata, 2007). In the U.S., national organisations such as Coca-Cola, Campbell Soup and Heinz led this phase (Son, 2015).

The segmentation phase (1950 to current)

Throughout this phase, organisations used television, internet and radio advertising in order to market its food products (Son, 2015). Through the use of these advertising methods, organisations were able to satisfy and understand particular geographic, demographic, psychographic and behavioural segments of the markets. This was achieved by displaying the different benefits and images of their food products (Murata, 2007).

Marketing Mix

The four components of food marketing are often called the "four Ps" of the marketing mix because they relate to product, price, promotion, and place. One reason food manufacturers receive the largest percentage of the retail food dollar is that they provide the most differentiating, value-added service. The money that manufacturers invest in developing, pricing, promotion, and placing their products helps differentiate a food product on the basis of both quality and brand-name recognition.

To understand how food marketing works as a whole, the "4 Ps" of the marketing mix must firstly be recognised. The 4Ps stand for product, price, promotion and place (Wongleedee, 2015). Organisations spend a large amount of revenue developing these aspects of the marketing mix in order to differentiate its food products, through quality and brand-name recognition (Pomeranz & Adler, 2015). Overall, the marketing mix can add value to a food organisation's product (Helm & Gritsch, 2014).

Product

In deciding what type of new food products a consumer would most prefer, a manufacturer can either try to develop a new food product or try to modify or extend an existing food. For example, a sweet, flavored yogurt drink would be a new product, but milk in a new flavor (such as chocolate strawberry) would be an extension of an existing product. There are three steps to both developing and extending: generate ideas, screen ideas for feasibility, and test ideas for appeal. Only after these steps will a food product make it to national market. Of one hundred new food product ideas that are considered, only six make it to a supermarket shelf.

The food industry faces numerous marketing decisions. Money can be invested in brand building (through advertising and other forms of promotion) to increase either quantity demanded or the price consumers are willing to pay for a product. Coca Cola, for example, spends a great deal of money both on perfecting its formula and on promoting the brand. This allows Coke to charge more for its product than can makers of regional and smaller brands. Manufacturers may be able to leverage their existing brand names by developing new product lines. For example, Heinz started out as a brand for pickles but branched out into ketchup. Some brand extensions may involve a risk of damage to the original brand if the quality is not good enough. Coca Cola, for example, refused to apply the Coke name to a diet drink back when artificial sweeteners had a significantly less attractive taste. Coke created Tab Cola, but only when aspartame (NutraSweet) was approved for use in soft drinks did Coca Cola come out with a Diet Coke. Manufacturers that have invested a great deal of money in brands may have developed a certain level of consumer brand loyalty—that is, a tendency for consumers to continue to buy a preferred brand even when an attractive offer is made by competitors. For loyalty to be present, it is not enough to merely observe that the consumer buys the same brand consistently. The consumer, to be brand loyal, must be able to actively resist promotional efforts by competitors. A brand loyal consumer will continue to buy the preferred brand even if a competing product is improved, offers a price promotion or premium, or receives preferred display space. Some consumers have multi-brand loyalty. Here, a consumer switches between a few preferred brands. The consumer may either alternate for variety or may, as a rule of thumb, buy whichever one of the preferred brands is on sale. This consumer, however, would not switch to other brands on sale. Brand loyalty is, of course, a matter of degree. Some consumers will not switch for a moderate discount, but would switch for a large one or will occasionally buy another brand for convenience or variety.

The product of the marketing mix refers to the goods and/or services that the organisation will offer to the consumer (Wongleedee, 2015). An organisation can achieve this by either creating a new food product, or by modifying or improving an existing food product (Pomeranz & Adler, 2015). For example, an organic almond yoghurt drink would be considered a new product, whereas a chocolate flavoured milk drink would be an extension of an existing product (Helm & Gritsch, 2014). The three steps to develop and extend a food product include generating ideas, analysing the ideas for feasibility and testing ideas for demand. Once these steps have successfully been completed, the food product can then be manufactured to the food market (Pomeranz & Adler, 2015).

Price

In profitably pricing the food, the manufacturer must keep in mind that the retailer adds approximately 50 percent to the price of a wholesale product. For example, a frozen food sold in a retail store for $4.50 generates an income of $3.00 for the manufacturer. This money has to pay for the cost of producing, packaging, shipping, storing, and selling the product.

Price encompasses the amount of money paid by the consumer in order to purchase the food product (Wongleedee, 2015). When pricing the food products, the manufacturer must bear in mind that the retailer will add a particular percentage to the price on the wholesale product. This percentage amount differs globally (Pomeranz & Adler, 2015). The percentage is used to pay for the cost of producing, packaging, shipping, storing and selling the food product. For example, the purchasing of a food product in a supermarket selling for $3.50 generates an income of $2.20 for the manufacturer (Wongleedee, 2015).

Promotion

Promoting a food to consumers is done out of store, in store, and on package. Advertisements on television and in magazines are attempts to persuade consumers to think favorably about a product, so that they go to the store to purchase the product. In addition to advertising, promotions can also include Sunday newspaper ads that offer coupons such as cents-off and buy-one-get-one-free offers.

Promotion of the marketing mix is defined as the actions used to communicate a food product's features and benefits; therefore, persuading the consumer to purchase the product (Wongleedee, 2015). There are multiple avenues used to promote a food product to consumers. Some examples are out-of-store, in-store and on the packaging (Helm & Gritsch, 2014). Food advertisements on television are used to attract the consumer, persuading them to go to the store and purchase that food product. Additionally, promotions in magazines and newspapers may offer coupons for food products (Frechette, 2015).

Place

Place refers to the distribution and warehousing efforts necessary to move a food from the manufacturer to a location where a consumer can buy it. It can also refer to where the product is located in a retail outlet (e.g., the end of an aisle; the top, bottom, or middle shelf; in a special display case, etc.).

The food marketing system in the United States is an amazingly flexible one. Consumer focus helps marketers anticipate the demands of consumers, and production focus helps them respond to changes in the market. The result is a system that meets and influences the ever-changing demands of consumers.

Place refers to the activities that organisations go about in order to make its food product available to its consumers. This encompasses the distribution necessary to move a food product from the manufacturer to a location where it can be purchased by the consumer (Wongleedee, 2015). Product location in a store is also a definition of place in the marketing mix. For example, a particular place in an aisle, a shelf or a display in a supermarket (Helm & Gritsch, 2014).

Food Marketing Segmentation

In order to market its food products, an organisation must first understand whether its product will satisfy the consumer's needs better than competitors do (Noori, 2015). In order to achieve this, an organisation must understand the four types of segmentation.

Geographic

An organisation must understand where it is marketing its food products to in a geographical sense (Schlanger & Maas, 2013). Clarifying this will help an organisation to grasp which food products will satisfy the needs of a particular consumer culture (Adeigbe, Baldwin, Gallion, Grier & Ramirez, 2015). For example, researching whether the consumer lives in America or Asia; or, whether the consumer lives in a city or a rural area (Schlanger & Maas, 2013). By understanding these aspects, the organisation will be able to identify that a consumers' food needs differ depending on where they live (Pomeranz & Adler, 2015).

Demographic

A food organisation must understand the demographic segment that it will be marketing towards (David, Schramm-Kelin, Rank & Wanger, 2015). Factors that must be considered are a consumer's age, gender, education, social class, income, religion and ethnicity (Quinn & Dibb, 2010). All of these aspects can impact whether the consumer will prefer one food product over another (Pomeranz & Adler, 2015).

Psychographic

A food organisation must understand its consumer psychographic. Factors such as the lifestyles, personalities, opinions, activities and interests of its potential consumers must considered (Schlanger & Maas, 2013). Identifying these aspects can help an organisation to improve its food products (Pomeranz & Adler, 2015).

Behaviour

A food organisation must understand how its consumers may behave towards a food product (David, Schramm-Kelin, Rank & Wanger, 2015). For example, information researching the benefits sought, frequency of food purchase, attitude towards the food product and nutritional knowledge of the product are all beneficial (Noori, 2015).

Issues in Food Marketing

Childhood Obesity:

It has been shown through Frechette's (2015) studies that in recent years, as children and teenagers have become more exposed to technological advances, they have become more susceptible to unhealthy food marketing commercial messages from food organisations (Cairns, Angus, Hastings & Caraher, 2013). Harris, Pomeranz, Lobstein, & Brownell (2009) suggest that food marketers have been using child-targeting marketing practices in order to persuade children to eat unhealthy or poorly suitable foods. Organisations achieve this through direct and indirect marketing tactics on television adverts, games, social media and food packaging (Freeman, Kelly, Baur, Chapman, Chapman, Gill & King, 2014). These tactics have an explicit effect on children's consumption patterns, diet-related health, nutritional knowledge, purchase behaviour and preferences (Vandevijvere & Swinburn, 2015). Overall, it has been found that food marketing is one of the leading contributors to an increase in childhood obesity (Schor & Ford, 2007). This is increasingly becoming a global issue (Chandon & Wansink, 2012).

In order to prevent the current unhealthy food marketing culture, Sacks, Mialon, Vandevijvere, Trevena, Snowdon, Crino & Swinburn (2015) believe that there are methods and policies that should be put into place by governments. Firstly, parents should be informed of the nutritional values of the foods that they are giving to their children (Grier, Mensinger, Huang, Kumanyika & Stettler, 2007). For example, an easy-to-read nutritional label on food packaging that provides the nutrient values and their definitions (Nikolova & Inman, 2015). This will create healthier food environments for families around the world (Sacks et al., 2015). Secondly, parents could restrict which advertisements that their children are exposed too (Newman & Oates, 2014). For example, parents could use ad-blocking applications or limit television watching time (Thaichon & Quach, 2016). Through the implementation of these strategies, governments can contribute in decreasing the rates of childhood obesity through food marketing (Hawkes, 2007).

Misleading Nutritional Information

In order to persuade consumers to buy food products, organisations may present misleading nutritional information on its food products (Schermel, Emrich, Arcand, Wong & L'abbé, 2013). It has become more prevalent that food organisations are not entirely transparent in regard to what is contained in its food products. For example, saturated fats, sodium and added sugars (Harris, LoDolce & Schwartz, 2015). Wording such as "less sugar", "fat free" and "all natural" lead consumers to believe that the foods they are consuming are healthy (Nikolova & Inman, 2015). Sacks et al. (2015) discusses that in order to prevent misleading food advertising, governments should implement policies regarding the placement of verified nutritional values on food packaging.

Criticism of Food Marketing

In recent years, food marketing has been criticised by professionals as being a major factor in contributing to childhood obesity. Nestle (2006) suggests that food marketing purposely targets children who are easily influenced at such a young age to eat high-sugar drinks and food with little nutritional benefit. The fact that areas of food marketing are linked to obesity is concerning, as well as how easy it is to market to children and affect their food choices. Television and print media still stand as the traditional communication channels for food marketing, but other popular approaches are also more prevalent these days. These include the Internet, toys, packaging, video games, blockbuster films, character licensing of children's toys and celebrity advertising (McGinnis, Appleton Gootman, & Kraak, 2006). The employment of these food marketing strategies are growing, and are said to be partly responsible for swelling rates of childhood obesity (Cartere, 2009).

Product placement in children's films and television shows gives food marketers more power to get children familiar with their brand and to directly interact with this market segment (Cartere, 2009). The power brands have through food marketing on television is significant because television audiences automatically are more enticed in an advertisement as it is playing in front of them, forming stronger predispositions for brands (Kline, 2010). Accusations come into play when this saturation happens as children are not equipped with adequate knowledge to make smart nutritional choices, and food marketing is therefore sometimes blamed for children's unhealthy lifestyles (Kline, 2010). Children are a fast expanding market segment, firstly because they yield influence over their parents buying, but also because they are future consumers themselves (Cartere, 2009). Food marketers capitalize on the fact most children trouble their parents for a product they have seen on television until they receive it, giving children high bargaining power. According to McGinnis et al. (2006), by the time children are two years old, the majority can identify brands in supermarkets and demand them by name. It has been argued that marketers also portray some foods to children as exclusive only to them, implying children know better about their nutrition than their parents. This has in turn seen trends of children disregarding parental advice and taking it upon themselves to pick and choose what they eat (Nestle, 2006).

Food marketers also use appealing packaging to attract children to their product through bright colours, including toys in schemes (McDonald's Happy Meals with a toy included is an example) and utilising famous television or film characters to spark interest (Cartere, 2009). In terms of packaging, brands will also change the size of products to entice children. Large companies have further been criticised of contributing to obesity through supplying schools with branded sponsorships and sports merchandise such as rugby balls that flaunt a company's logo (Cartere, 2009). Food marketers are criticised further than being responsible for child obesity rates, and are said to not have children's long-term physical wellbeing in mind when they aim to rapidly create brand name association among children (Frechette, 2015).

Purchasing Decisions in Food Marketing

Food marketing not only involves the marketing of products to consumers, but the reasons why consumers purchase these items and the factors influencing such choices (Sahay, Stough, Sohal, & Goyal, 2006). Demographics, values and attitudes, incentives, and price willingness to pay are all elements that drive buyer selection in the marketing of food.

Firstly, food marketers must be aware that the attitudes and values of their target market play a significant role in what they choose to buy. For example, in terms of the green brand market, consumers will first be environmentally conscious and therefore intend to buy such products. Once a product has been marketed to a consumer, they need to feel that they are contributing to the preservation of the environment to purchase something (Kapuge, 2016). Consumers that have concern for the environment tend to alter their behavior to be more environmentally friendly. Values and attitudes have been found to be the driving force of sustainable purchasing behavior (Kapuge, 2016). Marketers need to convey to their consumer market through information that some items have adverse effects on the environment. Brands need to also relay the values of food marketing to customers when communicating with them (Sahay et al., 2006).

In relation to knowledge and beliefs in variables affecting food marketing decisions, it has been suggested that someone's understanding of products helps anticipate their ecologically friendly actions (Nikolova & Inman, 2015). Because a persons awareness of information can cause them to make more informed or less informed decisions about products shows knowledge is a major factor in consumer buying choices. When strategists are food marketing, it is important to create binding relationships with potential customers through understanding their beliefs and awareness about the marketed product (Nikolova & Inman, 2015).

Demographics play a large role in determining the background of consumers and how food marketing operates. Such as age allows a brand to market its food towards certain age groups according to their wants. Education also comes into play because often education is associated with people buying better food (Sahay et al., 2006). Income alike sees if people have more disposable income to spend on good products. Gender further allows food marketers to target women especially because women have been found to do the majority of food shopping in families (Sahay et al., 2006).

Signage conveys a lot about a brand, for example what type of food they sell, the ingredients, where it was made and the cost of the products. Signage precisely communicates with the buyer and is an essential way to affect feelings and brain processes of a person (Remar, Campbell, & DiPietro, 2016). Findings show that certain digital signposting of a brand influences consumer perception and behavior towards a product, meaning food marketing strategies like this are very effective in brand positioning (Nikolova & Inman, 2015).

Lastly, consumer price willingness to pay is another tool used in food marketing to track how much a person is prepared to pay for a product. Marketers can take note of people's readiness to pay to see if they would indeed buy their product (Remar et al., 2016). Studies have shown that consumers are prepared to spend more on an item if it is environmentally friendly and portrays them as having sustainable behavior in society (Remar et al., 2016). Additionally, it has been explored that consumers will for example pay substantially more (almost 50%) for food produce that is said to be grown locally as opposed to grown in the wider country or imported (Remar et al., 2016). This is vital information when it comes to food marketing and helps companies make informed decisions about their food marketing strategies.

Food Marketing Via Social Media and its Influences

Social media platforms are an effective way for food marketers to enter or attract a market. 'Social media' refers to websites such as Facebook and Twitter, and apps such as Instagram and Facebook. These websites allow users to upload and share content such as information, pictures, opinions and recommendations with the touch of a button. Food companies use unique marketing techniques designed to connect with the consumer such as running competitions, holding free giveaways and creating interactive games and apps. (Freeman et al., 2014) The benefits of these techniques are distinctive to social media platforms, as they interact and engage with the consumer on a level that other traditional media platforms are not capable. (Freeman et al., 2014) The increase of social media platforms being used for food marketing are widely beneficial to food marketers, however the technological increase in society also allows consumers to block or delete advertisements while online and watching TV. AD Block and My Sky are prime examples of technological advances used by consumers to eliminate advertisements.

Supermarket catalogues, product placement and their involvement in food marketing

A study of supermarket catalogues distributed globally has also yielded information which supports the claim that food advertising is contributing to rising global obesity levels, especially in countries where the advertising is more prominent, consistent and persuasive. (Charlton, Kähkönen, Sacks & Cameron, 2015). The study concluded with the findings that supermarket catalogues have a major influence on the purchasing decisions and patterns of consumers. (Charlton et al., 2015) For example, countries with the most consistent, regular catalogues filled with unhealthy foods were among those testing highest for obesity as well. The study also noted that India and The Philippines were the only countries analysed that advertised a vast majority of healthy foods over unhealthy options. (Charlton et al., 2015).

Product placement is another food marketing technique that draws the consumer in and encourages them to purchase. Vending machines and pop-up sample booths are model examples of product placement techniques used by food marketers. Vending machines provide convenience to the consumer as they are usually stationed in areas with no larger food options nearby, such as university campuses and gyms. Consumers using vending machines will usually pay a slightly higher price for the item, and will usually not be concerned as the price difference accounts for the convenience provided to them.

Food Marketing Issues

There are no set worldwide food marketing laws or legislation, therefore countries have the option to adopt their own legislation with regards to food marketing standards. (Restrict Food Marketing, 2016). These standards are usually based on the values, culture and ethics of the particular country (Kaynak, 1999). Many countries worldwide have created laws in order to limit food marketing towards children with the goal of reducing rising obesity levels. (Restrict Food Marketing, 2016). Obesity has been proven to have increased significantly with a link to the increase of food marketing and advertising in society. (Dibb & Lobstein, 2005).

Following a 'call out' from the Worldwide Health Organisation in 2006, many countries adopted to change marketing laws and legislation in order to protect children from persuasive advertisements directly targeted at them. (Center for science in the public interest, 2007). These advertisements are strategically designed with special techniques in order to attract the children's attention.

The Chilean Government has taken some of the most drastic steps in 2012 by approving the 'Law of Nutritional Composition of Food and Advertising'. (Restrict Food Marketing, 2016). This law gives strict guidelines for food marketers, allowing no high in saturated fat, sodium and sugar products to be advertised to under 14 year olds or through television or websites with an audience or more than 20% children. (Restrict Food Marketing, 2016). Iran has also taken similar actions, having banned the advertisement of all soft drinks since 2004. Iran's Ministry of Health and Medical Education has also prepared a document in 2015 containing 24 items in which they hope to prohibit the advertising of in future. (Restrict Food Marketing, 2016). Dibb and Lobstein acknowledge that traditional media outlets only account for approximately 20% of a food marketer's budget, therefore children are still highly exposed to influential advertising in shopping malls and grocery stores, vending machines, sponsored toys, contests etc. (Dibb & Lobstein, 2005).

In addition to Iran and Chile's more extensive legislation's, over 30 countries worldwide have all adopted some form of legislation to protect children and reduce their exposure to advertisements. Some of these countries include Australia, Europe, Canada, Malaysia and Korea. (Center for science in the public interest, 2007). An example of the type of clauses in the legislation's include prohibiting the use of certain marketing techniques such as using cartoon characters, which can be used to access a child's mind and gain their attention. (Centre for science in the public interest, 2007). Using physiological techniques such as this can be ethically wrong as well as illegal in some countries.

Advertising to children at a young age is a well-established food marketing technique designed to encourage brand preference and holds many ethical dilemmas. Previous studies have concluded that children can recognize and mentally picture brand logos at the age of just six months old, and will verbally request brands at the age of 3 years. (As reported by The Campaign for a Commercial-Free Childhood). Marketing professor James McNeal acknowledged "The Drool Factor" – a study which recognized the fact that babies naturally stare down at their bibs while drooling to see where their drool lands. (Collins, 2013). Customizing baby bibs with brand logos has become an effective way for food marketers to imprint their brand into the child's lifestyle, targeting them at a vulnerable young age resulting in brand recognition from the child. (Collins, 2013) As a result, when the child is older they will continue to reciprocate warm, fond feelings towards the brand when encountering it in society. This is seen as unethical in many countries and is why organisations such as 'The Campaign for a Commercial-free Childhood' exist and what they aim to reduce.

Food Storage

Yup'ik elevated food cache (*qulvarvik*), Hooper Bay, Alaska, 1929. Photograph by Edward S. Curtis

U.S. Federal Emergency Management Agency (FEMA) food storage containers stacked on shipping pallets in Texas, 2008.

Food logistics is both a traditional domestic skill and is important industrially. The transport and storing of food as well as timely delivery to consumers is important in order to secure the procurement of food. Food is stored by almost every human society and by many animals. Storing of food has several main purposes:

- Storage of harvested and processed plant and animal food products for distribution to consumers
- Enabling a better balanced diet throughout the year
- Reducing kitchen waste by preserving unused or uneaten food for later use
- Preserving pantry food, such as spices or dry ingredients like rice and flour, for eventual use in cooking

- Preparedness for catastrophes, emergencies and periods of food scarcity or famine
- Religious reasons (Example: LDS Church leaders instruct church members to store food)
- Protection from animals or theft

History of Food Storage

Traditional Sami food storage structure to protect food from animals

Domestic Food Storage

Tupperware kitchen storage containers designed for a variety of uses.

The safe storage of food for home use should strictly adhere to guidelines set out by reliable sources, such as the United States Department of Agriculture. These guidelines have been thoroughly researched by scientists to determine the best methods for reducing the real threat of food poisoning from unsafe food storage. It is also important to maintain proper kitchen hygiene, to reduce risks of bacteria or virus growth and food poisoning. The common food poisoning illnesses include Listeriosis, Mycotoxicosis, Salmonellosis, E. coli, Staphylococcal food poisoning and Botulism. There are many other organisms that can also cause food poisoning.

There are also safety guidelines available for the correct methods of home canning of food. For example, there are specific boiling times that apply depending upon whether pressure canning or waterbath canning is being used in the process. These safety guidelines are intended to reduce the growth of mold and bacteria and the threat of potentially-fatal food poisoning.

Plastic storage containers can be used to store food.

Food Storage Safety

Freezers and Thawing food

Freezer temperature should be maintained below 0°F. Food should never be thawed at room temperature, this increases the risk of bacterial and fungal growth and accordingly the risk of food poisoning. Once thawed, food should be used and never refrozen. Frozen food should be thawed using the following methods:

- Microwave oven
- During cooking
- In cold water (place food in watertight, plastic bag; change water every 30 minutes)
- In the refrigerator

Throw out foods that have been warmer than 40 °F for more than 2 hours. If there is any doubt at all about the length of time the food has been defrosted at room temperature, it should be thrown out. Freezing does not destroy microbes present in food. Freezing at 0 °F does inactivate microbes (bacteria, yeasts and molds). However, once food has been thawed, these microbes can again become active. Microbes in thawed food can multiply to levels that can lead to foodborne illness. Thawed food should be handled according to the same guidelines as perishable fresh food.

Food frozen at 0°F and below is preserved indefinitely. However, the quality of the food will deteriorate if it is frozen over a lengthy period. The United States Department of Agriculture, Food Safety and Inspection Service publishes a chart showing the suggested freezer storage time for common foods.

Refrigeration

It is important to note that safe food storage using refrigeration requires adhering to temperature guidelines:

For safety, it is important to verify the temperature of the refrigerator. Refrigerators should be set to maintain a temperature of 40 °F or below. Some refrigerators have built-in thermometers to measure their internal temperature. For those refrigerators without this feature, keep an appliance thermometer in the refrigerator to monitor the temperature. This can be critical in the event of a power outage. When the power goes back on, if the refrigerator is still 40 °F, the food is safe. Foods held at temperatures above 40 °F for more than 2 hours should not be consumed. Appliance thermometers are specifically designed to provide accuracy at cold temperatures. Be sure refrigerator/freezer doors are closed tightly at all times. Don't open refrigerator/freezer doors more often than necessary and close them as soon as possible.

Storage Times for Refrigerated Food

The United States Department of Agriculture, Food Safety and Inspection Service publishes recommended storage times for refrigerated food.

Storing Oils and Fats

Oils and fats can begin to go rancid quickly when not stored safely. Rancid cooking oils and fats do not often smell rancid until well after they have spoiled. Oxygen, light and heat all contribute to cooking oils becoming rancid. The higher the level of polyunsaturated fat that an oil contains, the faster it spoils. The percentage of polyunsaturated fat in some common cooking oils is: safflower (74%); sunflower (66%); corn (60%); soybean (37%); peanut (32%); canola (29%); olive (8%).

To help preserve oils from rancidification, they should be refrigerated once opened. Opened, refrigerated cooking oils should be used within a few weeks, when some types begin to go rancid. Unopened oils can have a storage life of up to one year, but some types have a shorter shelf-life even when unopened (such as sesame and flaxseed).

Dry Storage of Foods

Vegetables

A large root cellar at the Oxon Hill Manor farm in Maryland

The guidelines vary for safe storage of vegetables under dry conditions (without refrigerating or freezing). This is because different vegetables have different characteristics, for example, tomatoes contain a lot of water, while root vegetables such as carrots and potatoes contain less. These factors, and many others, affect the amount of time that a vegetable can be kept in dry storage, as well as the temperature needed to preserve its usefulness. The following guideline shows the required dry storage conditions:

- Cool and dry: onion
- Cool and moist: root vegetable, potato, cabbage

- Warm and dry: winter squash, pumpkin, sweet potatoes, dried hot peppers

Many cultures have developed innovative ways of preserving vegetables so that they can be stored for several months between harvest seasons. Techniques include pickling, home canning, food dehydration, or storage in a root cellar.

Grain

Grain, which includes dry kitchen ingredients such as flour, rice, millet, couscous, cornmeal, and so on, can be stored in rigid sealed containers to prevent moisture contamination or insect or rodent infestation. For kitchen use, glass containers are the most traditional method. During the 20th century plastic containers were introduced for kitchen use. They are now sold in a vast variety of sizes and designs.

Metal cans are used (in the United States the smallest practical grain storage uses closed-top #10 metal cans). Storage in grain sacks is ineffective; mold and pests destroy a 25 kg cloth sack of grain in a year, even if stored off the ground in a dry area. On the ground or damp concrete, grain can spoil in as little as three days, and the grain might have to be dried before it can be milled. Food stored under unsuitable conditions should not be purchased or used because of risk of spoilage. To test whether grain is still good, it can be sprouted. If it sprouts, it is still good, but if not, it should not be eaten. It may take up to a week for grains to sprout. When in doubt about the safety of the food, throw it out as quickly as possible.

Spices and Herbs

Spices and herbs are today often sold prepackaged in a way that is convenient for pantry storage. The packaging has dual purposes of both storing and dispensing the spices or herbs. They are sold in small glass or plastic containers or resealable plastic packaging. When spices or herbs are homegrown or bought in bulk, they can be stored at home in glass or plastic containers. They can be stored for extended periods, in some cases for years. However, after 6 months to a year, spices and herbs will gradually lose their flavour as oils they contain will slowly evaporate during storage.

Spices and herbs can be preserved in vinegar for short periods of up to a month, creating a flavoured vinegar.

Alternative methods for preserving herbs include freezing in water or unsalted butter. Herbs can be chopped and added to water in an ice cube tray. After freezing, the ice cubes are emptied into a plastic freezer bag for storing in the freezer. Herbs also can be stirred into a bowl with unsalted butter, then spread on wax paper and rolled into a cylinder shape. The wax paper roll containing the butter and herbs is then stored in a freezer, and can be cut off in the desired amount for cooking. Using either of these techniques, the herbs should be used within a year.

Meat

Unpreserved meat has only a relatively short life in storage. Perishable meats should be refrigerated, frozen, dried promptly or cured. Storage of fresh meats is a complex discipline that affects the costs, storage life and eating quality of the meat, and the appropriate techniques vary with the kind of meat and the particular requirements. For example, dry ageing techniques are sometimes used to tenderize gourmet meats by hanging them in carefully controlled environments for up to 21 days, while game animals of various kinds may be hung after shooting. Details depend on personal tastes and local traditions. Modern techniques of preparing meat for storage vary with the type of meat and special requirements of tenderness, flavour, hygiene, and economy.

Semi-dried meats like salamis and country style hams are processed first with salt, smoke, sugar, acid, or other "cures" then hung in cool dry storage for extended periods, sometimes exceeding a year. Some of the materials added during the curing of meats serve to reduce the risks of food poisoning from anaerobic bacteria such as species of Clostridium that release botulinum toxin that can cause botulism. Typical ingredients of curing agents that inhibit anaerobic bacteria include nitrates. Such salts are dangerously poisonous in their own right and must be added in carefully controlled quantities and according to proper techniques. Their proper use has however saved many lives and much food spoilage.

Like the semi-dried meats, most salted, smoked, and simply-dried meats of different kinds that once were staples in particular regions, now are largely luxury snacks or garnishes; examples include jerky, biltong, and varieties of pemmican, but ham and bacon for instance, still are staples in many communities.

Food Rotation

Food rotation is important to preserve freshness. When food is rotated, the food that has been in storage the longest is used first. As food is used, new food is added to the pantry to replace it; the essential rationale is to use the oldest food as soon as possible so that nothing is in storage too long and becomes unsafe to eat. Labelling food with paper labels on the storage container, marking the date that the container is placed in storage, can make this practice simpler. The best way to rotate food storage is to prepare meals with stored food on a daily basis.

For Emergency Preparation

Guides for surviving emergency conditions in many parts of the world recommend maintaining a store of essential foods; typically water, cereals, oil, dried milk, and protein rich foods such as beans, lentils, tinned meat and fish. A food storage calculator can be used to help determine how much of these staple foods a person would need to store in order to sustain life for one full year. In addition to storing the basic food items many people choose to supplement their food storage with frozen or preserved garden-grown fruits and vegetables and freeze-dried or canned produce. An unvarying diet of staple foods prepared in the same manner can cause appetite exhaustion, leading to less caloric intake. Another benefit to having a basic supply of food storage in the home is for the potential cost savings. Costs of dry bulk foods (before preparation) are often considerably less than convenience and fresh foods purchased at local markets or supermarkets. There is a significant market in convenience foods for campers, such as dehydrated food products.

Commercial Food Logistics

Grain and beans are stored in tall grain elevators, almost always at a rail head near the point of production. The grain is shipped to a final user in hopper cars. In the former Soviet Union, where harvest was poorly controlled, grain was often irradiated at the point of production to suppress mold and insects. In the U.S., threshing and drying is performed in the field, and transport is nearly sterile and in large containers that effectively suppresses pest access, which eliminates the need for irradiation. At any given time, the U.S. usually has about two weeks worth of stored grains for the population.

Silos connected to a grain elevator on a farm in Israel.

Fresh fruits and vegetables are sometimes packed in plastic packages and cups for fresh premium markets, or placed in large plastic tubs for sauce and soup processors. Fruits and vegetables are usually refrigerated at the earliest possible moment, and even so have a shelf life of two weeks or less.

In the United States, livestock is usually transported live, slaughtered at a major distribution point, hung and transported for two days to a week in refrigerated rail cars, and then butchered and sold locally. Before refrigerated rail cars, meat had to be transported live, and this placed its cost so high that only farmers and the wealthy could afford it every day. In Europe much meat is transported live and slaughtered close to the point of sale. In much of Africa and Asia most meat is for local populations is raised, slaughtered and eaten locally, which is believed to be less stressful for the animals involved and minimizes meat storage needs. In Australia and New Zealand, where a large proportion of meat production is for export, meat enters the cold chain early, being stored in large freezer plants before being shipped overseas in freezer ships.

References

- Marotz, Lynn R. (2008). Health, Safety, and Nutrition for the Young Child. Wadsworth Publishing. pp. 491–2. ISBN 978-1-4283-2070-3.
- Marotz, Lynn R. (2008). Health, Safety, and Nutrition for the Young Child. Wadsworth Publishing. p. 482. ISBN 978-1-4283-2070-3.
- Mitchell, Deborah (2004). Safe foods: the A-to-Z guide to the most wholesome foods for you and your family. Penguin. pp. Ch. 15. ISBN 978-1-101-21015-4.
- Editors of Creative Publishing. Dressing & Cooking Wild Game. Publisher: Creative Publishing international 1999 ISBN 978-0865731080
- "Solutions for Your Toughest Mixing Applications in Food De-agglomerating" (PDF). Silverson Machines Application Report. Retrieved 2012-02-29.
- "Fact Sheet: Freezing and Food Safety". United States Department of Agriculture, Food Safety and Inspection Service. June 3, 2010. Retrieved November 8, 2011.

4

Food Chemistry: An Integrated Study

Food chemistry studies the chemical reactions that take place in food and analyses the interaction of the various components of food. It also studies any change that takes place in food composition like the coagulation of milk into cheese or the browning of cut apples. This section provides valuable information on food coloring, food additives or preservatives, flavors and the nutrient content of food. The chapter elucidates the crucial theories and principles of food chemistry.

Food Chemistry

Food chemistry is the study of chemical processes and interactions of all biological and non-biological components of foods. The biological substances include such items as meat, poultry, lettuce, beer, and milk as examples. It is similar to biochemistry in its main components such as carbohydrates, lipids, and protein, but it also includes areas such as water, vitamins, minerals, enzymes, food additives, flavors, and colors. This discipline also encompasses how products change under certain food processing techniques and ways either to enhance or to prevent them from happening. An example of enhancing a process would be to encourage fermentation of dairy products with microorganisms that convert lactose to lactic acid; an example of preventing a process would be stopping the browning on the surface of freshly cut Red Delicious apples using lemon juice or other acidulated water.

History of Food Chemistry

The scientific approach to food and nutrition arose with attention to agricultural chemistry in the works of J. G. Wallerius, Humphry Davy, and others. For example, Davy published *Elements of Agricultural Chemistry, in a Course of Lectures for the Board of Agriculture* (1813) in the United Kingdom which would serve as a foundation for the profession worldwide, going into a fifth edition. Earlier work included that by Carl Wilhelm Scheele who isolated malic acid from apples in 1785.

In 1874 the Society of Public Analysts was formed, with the aim of applying analytical methods to the benefit of the public. Its early experiments were based on bread, milk and wine.

It was also out of concern for the quality of the food supply, mainly food adulteration and contamination issues that would first stem from intentional contamination to later with chemical food additives by the 1950s. The development of colleges and universities worldwide, most notably in the United States, would expand food chemistry as well with research of the dietary substances, most notably the Single-grain experiment during 1907-11. Additional research by Harvey W. Wiley at the United States Department of Agriculture during the late 19th century would play a key factor in the creation of the United States Food and Drug Administration in 1906. The American Chemical Society would establish their Agricultural and Food Chemistry Division in 1908 while the Institute of Food Technologists would establish their Food Chemistry Division in 1995.

Food chemistry concepts are often drawn from rheology, theories of transport phenomena, physical and chemical thermodynamics, chemical bonds and interaction forces, quantum mechanics and reaction kinetics, biopolymer science, colloidal interactions, nucleation, glass transitions and freezing/disordered or noncrystalline solids, and thus has Food Physical Chemistry as a foundation area.

Water in Food Systems

A major component of food is water, which can encompass anywhere from 50% in meat products to 95% in lettuce, cabbage, and tomato products. It is also an excellent place for bacterial growth and food spoilage if it is not properly processed. One way this is measured in food is by water activity which is very important in the shelf life of many foods during processing. One of the keys to food preservation in most instances is reduce the amount of water or alter the water's characteristics to enhance shelf-life. Such methods include dehydration, freezing, and refrigeration This field encompasses the *"physiochemical principles of the reactions and conversions that occur during the manufacture, handling, and storage of foods"*.

Carbohydrates

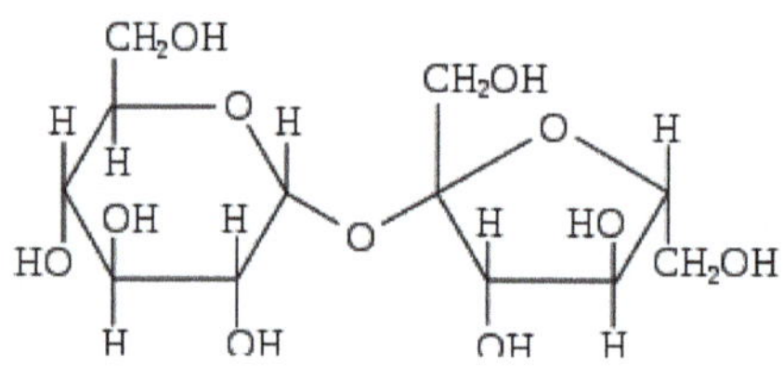

Sucrose: ordinary table sugar and probably the most familiar carbohydrate.

Comprising 75% of the biological world and 80% of all food intake for human consumption, the most common known human carbohydrate is Sucrose. The simplest version of a carbohydrate is a monosaccharide which possesses the properties of carbon, hydrogen, and oxygen in a 1:2:1 ratio under a general formula of $C_nH_{2n}O_n$ where n is a minimum of 3. Glucose is an example of a monosaccharide as is fructose. Combine them in the picture shown to the right and you have sucrose, one of the more common sugar products around.

A chain of monosaccharides form to make a polysaccharide. Such polysaccharides include pectin, dextran, agar, and xanthan.

Sugar content is commonly measured in degrees brix.

Lipids

The term lipid comprises a diverse range of molecules and to some extent is a catchall for relatively water-insoluble or nonpolar compounds of biological origin, including waxes, fatty acids (including essential fatty acids), fatty-acid derived phospholipids, sphingolipids, glycolipids and terpenoids, such as retinoids and steroids. Some lipids are linear aliphatic molecules, while others have ring structures. Some are aromatic, while others are not. Some are flexible, while others are rigid.

Most lipids have some polar character in addition to being largely nonpolar. Generally, the bulk of their structure is nonpolar or hydrophobic ("water-fearing"), meaning that it does not interact well with polar solvents like water. Another part of their structure is polar or hydrophilic ("water-loving") and will tend to associate with polar solvents like water. This makes them amphiphilic molecules (having both hydrophobic and hydrophilic portions). In the case of cholesterol, the polar group is a mere -OH (hydroxyl or alcohol).

Lipids in food include the oils of such grains as corn, soybean, from animal fats, and are parts of many foods such as milk, cheese, and meat. They also act as vitamin carriers as well.

Food Proteins

Proteins compose over 50% of the dry weight of an average living cell and are very complex macromolecules. They also play a fundamental role in the structure and function of cells. Consisting mainly of carbon, nitrogen, hydrogen, oxygen, and some sulfur, they also may contain iron, copper, phosphorus, or zinc.

In food, proteins are essential for growth and survival and vary depending upon a person's age and physiology (e.g., pregnancy). Protein is commonly obtained from animal sources: eggs, milk, and meat. Nuts, grains and legumes provide vegetable sources of protein, and protein combining of vegetable sources is used to achieve complete protein nutritional quotas from vegetables.

Protein sensitivity as food allergy is detected with the ELISA test.

Enzymes

Enzymes are biochemical catalysts used in converting processes from one substance to another. They are also involved in reducing the amount of time and energy required to complete a chemical process. Many aspects of the food industry use catalysts, including baking, brewing, dairy, and fruit juices, to make cheese, beer, and bread.

Vitamins

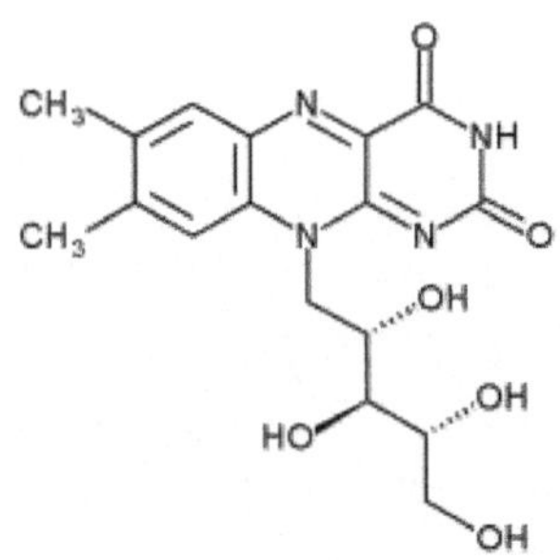

Riboflavin (Vitamin B_2), water-soluble.

Vitamins are nutrients required in small amounts for essential metabolic reactions in the body. These are broken down in nutrition as either water-soluble (Vitamin C) or fat-soluble (Vitamin E). An adequate supply of vitamins can prevent diseases such as beriberi, anemia, and scurvy while an overdose of vitamins can produce nausea and vomiting or even death.

Minerals

Dietary minerals in foods are large and diverse with many required to function while other trace elements can be hazardous if consumed in excessive amounts. Bulk minerals with a Reference Daily Intake (RDI, formerly Recommended Daily Allowance (RDA)) of more than 200 mg/day are calcium, magnesium, and potassium while important trace minerals (RDI less than 200 mg/day) are copper, iron, and zinc. These are found in many foods, but can also be taken in dietary supplements.

Color

Food coloring is added to change the color of any food substance. It is mainly for sensory analysis purposes. It can be used to simulate the natural color of a product as perceived by the customer, such as red dye (like FD&C Red No.40 Allura Red AC) to ketchup or to add unnatural colors to a product like Kellogg's Froot Loops. Caramel is a natural food dye; the industrial form, caramel coloring, is the most widely used food coloring and is found in foods from soft drinks to soy sauce, bread, and pickles.

Flavors

Flavor in food is important in how food smells and tastes to the consumer, especially in sensory analysis. Some of these products occur naturally like salt and sugar, but flavor chemists (called a "flavorist") develop many of these flavors for food products. Such artificial flavors include methyl salicylate which creates the wintergreen odor and lactic acid which gives milk a tart taste.

Food Additives

Food additives are substances added to food for preserving flavors, or improving taste or appearance. The processes are as old as adding vinegar for pickling or as an emulsifier for emulsion mixtures like mayonnaise. These are generally listed by "E number" in the European Union or GRAS ("generally recognized as safe") by the United States Food and Drug Administration.

Food Additive

Food additives are substances added to food to preserve flavor or enhance its taste and appearance. Some additives have been used for centuries; for example, preserving food by pickling (with vinegar), salting, as with bacon, preserving sweets or using sulfur dioxide as with wines. With the advent of processed foods in the second half of the twentieth century, many more additives have been introduced, of both natural and artificial origin.

Numbering

To regulate these additives, and inform consumers, each additive is assigned a unique number, termed as "E numbers", which is used in Europe for all approved additives. This numbering scheme has now been adopted and extended by the *Codex Alimentarius* Commission to internationally identify all additives, regardless of whether they are approved for use.

E numbers are all prefixed by "E", but countries outside Europe use only the number, whether the additive is approved in Europe or not. For example, acetic acid is written as E260 on products sold in Europe, but is simply known as additive 260 in some countries. Additive 103, alkanet, is not approved for use in Europe so does not have an E number, although it is approved for use in Australia and New Zealand. Since 1987, Australia has had an approved system of labelling for additives in packaged foods. Each food additive has to be named or numbered. The numbers are the same as in Europe, but without the prefix "E".

The United States Food and Drug Administration (FDA) lists these items as "generally recognized as safe" (GRAS); they are listed under both their Chemical Abstracts Service number and FDA regulation under the United States Code of Federal Regulations.

Categories

Food additives can be divided into several groups, although there is some overlap because some additives exert more than one effect. For example, salt is both a preservative as well as a flavor.

Acidulents

> confer sour or acid taste. Common acidulents include vinegar, citric acid, tartaric acid, malic acid, fumaric acid, and lactic acid.

Acidity regulators

> Acidity regulators are used to control the pH of foods for stability or to affect activity of enzymes.

Anticaking agents

> Anticaking agents keep powders such as milk powder from caking or sticking.

Antifoaming and foaming agents

> Antifoaming agents reduce or prevent foaming in foods. Foaming agents do the reverse.

Antioxidants

> Antioxidants such as vitamin C are preservatives by inhibiting the degradation of food by oxygen.

Bulking agents

> Bulking agents such as starch are additives that increase the bulk of a food without affecting its taste.

Food coloring

> Colorings are added to food to replace colors lost during preparation or to make food look more attractive.

Color retention agents

In contrast to colorings, color retention agents are used to preserve a food's existing color.

Emulsifiers

Emulsifiers allow water and oils to remain mixed together in an emulsion, as in mayonnaise, ice cream, and homogenized milk.

Flavors

Flavors are additives that give food a particular taste or smell, and may be derived from natural ingredients or created artificially.

Flavor enhancers

Flavor enhancers enhance a food's existing flavors. A popular example is monosodium glutamate. Some flavor enhancers have little flavor independent of the food.

Flour treatment agents

Flour treatment agents are added to flour to improve its color or its use in baking.

Glazing agents

Glazing agents provide a shiny appearance or protective coating to foods.

Humectants

Humectants prevent foods from drying out.

Tracer gas

Tracer gas allow for package integrity testing to prevent foods from being exposed to atmosphere, thus guaranteeing shelf life.

Preservatives

Preservatives prevent or inhibit spoilage of food due to fungi, bacteria and other microorganisms.

Stabilizers

Stabilizers, thickeners and gelling agents, like agar or pectin (used in jam for example) give foods a firmer texture. While they are not true emulsifiers, they help to stabilize emulsions.

Sweeteners

Sweeteners are added to foods for flavoring. Sweeteners other than sugar are added to keep the food energy (calories) low, or because they have beneficial effects for diabetes mellitus and tooth decay and diarrhea.

Thickeners

Thickening agents are substances which, when added to the mixture, increase its viscosity without substantially modifying its other properties.

Safety and Regulation

With the increasing use of processed foods since the 19th century, food additives are more widely used. Many countries regulate their use. For example, boric acid was widely used as a food preservative from the 1870s to the 1920s, but was banned after World War I due to its toxicity, as demonstrated in animal and human studies. During World War II, the urgent need for cheap, available food preservatives led to it being used again, but it was finally banned in the 1950s. Such cases led to a general mistrust of food additives, and an application of the precautionary principle led to the conclusion that only additives that are known to be safe should be used in foods. In the United States, this led to the adoption of the Delaney clause, an amendment to the Federal Food, Drug, and Cosmetic Act of 1938, stating that no carcinogenic substances may be used as food additives. However, after the banning of cyclamates in the United States and Britain in 1969, saccharin, the only remaining legal artificial sweetener at the time, was found to cause cancer in rats. Widespread public outcry in the United States, partly communicated to Congress by postage-paid postcards supplied in the packaging of sweetened soft drinks, led to the retention of saccharin, despite its violation of the Delaney clause. However, in 2000, saccharin was later found to only be carcinogenic to rats due to their unique urine chemistry.

Hyperactivity

Periodically, concerns have been expressed about a linkage between additives and hyperactivty, however "no clear evidence of ADHD was provided".

In 2007, Food Standards Australia New Zealand published an official shoppers' guidance with which the concerns of food additives and their labeling are mediated. In the EU it can take 10 years or more to obtain approval for a new food additive. This includes five years of safety testing, followed by two years for evaluation by the European Food Safety Authority and another three years before the additive receives an EU-wide approval for use in every country in the European Union. Apart from testing and analyzing food products during the whole production process to ensure safety and compliance with regulatory standards, Trading Standards officers (in the UK) protect the public from any illegal use or potentially dangerous mis-use of food additives by performing random testing of food products.

There has been significant controversy associated with the risks and benefits of food additives. Natural additives may be similarly harmful or be the cause of allergic reactions in certain individuals. For example, safrole was used to flavor root beer until it was shown to be carcinogenic. Due to the application of the Delaney clause, it may not be added to foods, even though it occurs naturally in sassafras and sweet basil.

Micronutrients

A subset of food additives, micronutrients added in food fortification processes preserve nutrient value by providing vitamins and minerals to foods such as flour, cereal, margarine and milk which normally would not retain such high levels. Preservatives also reduce spoilage from sources such as air, bacteria, fungi, and yeast.

Standardization of its Derived Products

ISO has published a series of standards regarding the topic and these standards are covered by ICS 67.220.

Science

Many food additives absorb radiation in the ultraviolet and / or visible region of the spectrum. This absorbance can be used to determine the concentration of an additive in a sample using external calibration. However, additives may occur together and the absorbance by one could interfere with the absorbance of another. A prior separation stage is necessary and the additives are first separated by high-pressure liquid chromatography (HPLC) and then determined on-line using a UV and/ or visible light detector.

Food Coloring

The addition of food coloring, such as beta-carotene, gives margarine its yellow color.

Food coloring, or color additive, is any dye, pigment or substance that imparts color when it is added to food or drink. They come in many forms consisting of liquids, powders, gels, and pastes. Food coloring is used both in commercial food production and in domestic cooking. Due to its safety and general availability, food coloring is also used in a variety of non-food applications including cosmetics, pharmaceuticals, home craft projects, and medical devices.

Purpose of Food Coloring

People associate certain colors with certain flavors, and the color of food can influence the perceived flavor in anything from candy to wine. Sometimes the aim is to simulate a color that is perceived by the consumer as natural, such as adding red coloring to glacé cherries (which would otherwise be beige), but sometimes it is for effect, like the green ketchup that Heinz launched in 1999. Color additives are used in foods for many reasons including:

- Offset color loss due to exposure to light, air, temperature extremes, moisture and storage conditions
- Correct natural variations in color
- Enhance colors that occur naturally
- Provide color to colorless and "fun" foods
- Make food more attractive and appetizing, and informative
- Allow consumers to identify products on sight, like candy flavors or medicine dosages
- Food colorants, synthetic

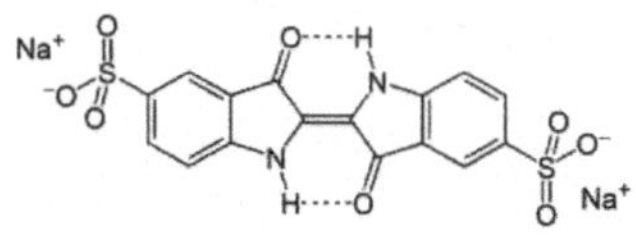

Indigo carmine, which is blue.

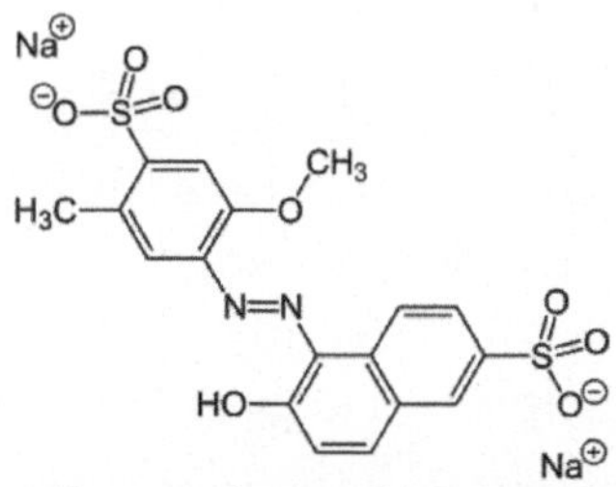

Allura Red AC, which is red.

Regulation

Past Use and Regulation History

The addition of colorants to foods is thought to have occurred in Egyptian cities as early as 1500 BC, when candy makers added natural extracts and wine to improve the products' appearance. During the Middle Ages, the economy in the European countries was based on agriculture, and the peasants were accustomed to producing their own food locally or trading within the village communities. Under feudalism, aesthetic aspects were not considered, at least not by the vast majority of the generally very poor population. This situation changed with urbanization at the beginning of the Modern Age, when trade emerged—especially the import of precious spices and colors. One of the very first food laws, created in Augsburg, Germany, in 1531, concerned spices or colorants and required saffron counterfeiters to be burned.

With the onset of the industrial revolution, people became dependent on foods produced by others. These new urban dwellers demanded food at low cost. Analytical chemistry was still primitive and regulations few. The adulteration of foods flourished. Heavy metal and other inorganic element-containing compounds turned out to be cheap and suitable to "restore" the color of watered-down milk and other foodstuffs, some more lurid examples being:

- Red lead (Pb_3O_4) and vermillion (HgS) were routinely used to color cheese and confectionery.

- Copper arsenite ($CuHAsO_3$) was used to recolor used tea leaves for resale. It also caused two deaths when used to color a dessert in 1860.

Sellers at the time offered more than 80 artificial coloring agents, some invented for dyeing textiles, not foods.

...Thus, with potted meat, fish and sauces taken at breakfast he would consume more or less Armenian bole, red lead, or even bisulphuret of mercury. At dinner with his curry or cayenne he would run the chance of a second dose of lead or mercury; with pickles, bottled fruit and vegetables he would be nearly sure to have copper administrated to him; and while he partook of bon-bons at dessert, there was no telling of the number of poisonous pigments he might consume. Again his tea if mixed or green, he would certainly not escape without the administration of a little Prussian blue...

Many color additives had never been tested for toxicity or other adverse effects. Historical records show that injuries, even deaths, resulted from tainted colorants. In 1851, about 200 people were poisoned in England, 17 of them fatally, directly as a result of eating adulterated lozenges. In 1856, mauveine, the first synthetic color, was developed by Sir William Henry Perkin and by the turn of the century, unmonitored color additives had spread through Europe and the United States in all sorts of popular foods, including ketchup, mustard, jellies, and wine.

Concerns over food safety led to numerous regulations throughout the world,. German food regulations released in 1882 stipulated the exclusion of dangerous minerals such as arsenic, copper, chromium, lead, mercury and zinc, which were frequently used as ingredients in colorants. In contrast to today, these first laws followed the principle of a negative listing (substances not allowed for use); they were already driven by the main principles of today's food regulations all over the world, since all of these regulations follow the same goal: the protection of consumers from toxic substances and from fraud. In the United States, the Pure Food and Drug Act of 1906 reduced the permitted list of synthetic colors from 700 down to seven. Even with updated food laws, adulteration continued for many years and this, together with more recent adverse press comments on food colors and health, has continued to contribute to consumer concern about color addition to foodstuffs.

In the 20th century, the improvement of chemical analysis and the development of trials to identify the toxic features of substances added to foods led to the replacement of the negative lists by lists of substances allowed to be used for the production and the improvement of foods. This principle is called a positive listing, and almost all recent legislations are based on it. Positive listing implies that substances meant for human consumption have been tested for their safety, and that they have to meet specified purity criteria prior to their approval by the corresponding authorities. In 1962, the first EU directive (62/2645/EEC) approved 36 colorants, of which 20 were naturally derived and 16 were synthetic. This directive did not list which food products the colorants could or could not be used in. At that time, each member state could designate where certain colors could and could not be used. In Germany, for example, quinoline yellow was allowed in puddings and desserts, but tartrazine was not. The reverse was true in France. This was updated in 1989 with 89/107/EEC, which concerned food additives authorized for use in foodstuffs.

Current Regulation

While naturally derived colors are not required to be certified by a number of regulatory bodies throughout the world (including the U.S. FDA), they still need to be approved for use in that country. The FDA lists "color additives exempt from certification" for food in subpart A of the Code of Federal Regulations - Title 21 Part 73. However, this list contains substances which may have synthetic origins, such as nature identical beta-carotene. FDA's permitted colors are classified as subject to certification or exempt from certification, both of which are subject to rigorous safety standards prior to their approval and listing for use in foods.

- Certified colors are synthetically produced and are used widely because they impart an intense, uniform color, are less expensive, and blend more easily to create a variety of hues. There are nine certified color additives approved for use in the United States. Certified food colors generally do not add undesirable flavors to foods.
- Colors that are exempt from certification include pigments derived from natural sources such as vegetables, minerals or animals. Nature derived color additives are typically more expensive than certified colors and may add unintended flavors to foods. Examples of exempt colors include annatto, beet extract, caramel, beta-carotene, turmeric and grape skin extract.

Food colorings are tested for safety by various bodies around the world and sometimes different bodies have different views on food color safety. In the United States, FD&C numbers (which indicate that the FDA has approved the colorant for use in foods, drugs and cosmetics) are given to approved synthetic food dyes that do not exist in nature, while in the European Union, E numbers are used for all additives, both synthetic and natural, that are approved in food applications. The food colors are known by E numbers that begin with a 1, such as E100 (turmeric) or E161b (lutein). The safety of food colors and other food additives in the EU is evaluated by the European Food Safety Authority. Color Directive 94/36/EC, enacted by the European Commission in 1994, outlines permitted natural and artificial colors with their approved applications and limits in different foodstuffs. This is binding to all member countries of the EU. Any changes have to be implemented into their national laws within a given time frame. In non-EU member states, food additives are regulated by their national authorities, which usually, but not in all cases, try to harmonize with the laws adopted by the EU. Most other countries have their own regulations and list of food colors which can be used in various applications, including maximum daily intake limits.

Global Harmonization

Global trade requires harmonization of food regulations on a world-wide basis in order to abolish barriers of trade and to ensure that the economic and nutritional demands of all nations are considered. Since the beginning of the 1960s, JECFA has played an important role in the development of international standards for food additives, not only by its toxicological assessments, which are continuously published by the WHO in a "Technical Report Series", but furthermore by elaborating appropriate purity criteria, which are laid down in the two volumes of the "Compendium of Food Additive Specifications" and their supplements. These specifications are not legally binding but very often serve as a guiding principle, especially in countries where no scientific expert committees have been established.

In order to further regulate the use of these evaluated additives, in 1962 the WHO and FAO created an international commission, the Codex Alimentarius, which is composed of authorities, food industry associations and consumer groups from all over the world. Within the Codex organization, the Codex Committee for Food Additives and Contaminants is responsible for working out recommendations for the application of food additives, the General Standard for Food Additives. In the light of the World Trade Organizations General Agreement on Tariffs and Trade (GATT), the Codex Standard, although not legally binding, influences food color regulations all over the world.

Artificial Coloring

As the 1900s began, a host of synthetic dyes and pigments became available. Originally, these were dubbed `coal-tar' colors because the starting materials were obtained from bituminous coal. Many synthesized dyes were easier and less costly to produce and were superior in coloring properties when compared to naturally derived alternatives of the time.

Food coloring spreading on a thin water film in the International Space Station

Also known as "azo-dyes", synthetic colors are generally produced via a two-step chemical synthesis. The first step forms a diazo compound from the reaction of aromatic amines generally formed from nitrosamine and a diazonium compound. The second step couples these diazo compounds with various reactive aromatic hydrocarbons. Due to the π-electrons across the two aromatic sections and the azo-groups, a conjugated system exists that is able to absorb light of specific wavelengths, leading to the color of the compounds (this principle also applies to naturally derived pigments as well).

The attractiveness of the synthetic dyes is that their color, lipophilicity, and other attributes can be engineered by the design of the specific molecule. The color of the dyes can be controlled selecting the number of azo-groups and various substituents. Yellow shades are achieved by using acetoacetanilide and heterocyclic compounds. Red colors result from the reaction between an aniline derivative (diazo) with a naphthol derivate. A blue results from replacing the aniline derivate with a benzidine derivate. The pair indigo and indigo carmine exhibit the same color, but the former is soluble in lipids, and the latter is water soluble because it has been fitted with sulfonate functional groups.

In The United States

Only seven dyes were initially approved under the U.S. Pure Food and Drug Act of 1906, but several have been delisted and replacements have been found. Some of the food colorings have the abbreviation "FCF" in their names. This stands for "For Coloring Food" (US) or "For Colouring of Food" (UK).

In the United States, manufacturers must get FDA approval for every batch of certified colors produced. FDA personnel evaluate the color's physical appearance and chemically analyze it. Tests include purity (total color content), moisture, residual salts, unreacted intermediates, colored impurities other than the main color (called subsidiary colors), any other specified impurities, and heavy metals (lead, arsenic, and mercury). The results are reviewed for compliance with the identity and specifications described in the listing regulation for the color additive. If the sample is found to meet these requirements, the FDA issues a certificate for the batch that identifies the color additive. For example, a batch of "tartrazine" becomes "FD&C Yellow No. 5."

Current Seven Generally Permitted

In the US, the following seven artificial colorings are generally permitted in food (the most common in bold) as of 2016. The lakes of these colorings are also permitted except the lake of Red No. 3.

- FD&C Blue No. 1 – Brilliant Blue FCF, E133 (blue shade)
- FD&C Blue No. 2 – Indigotine, E132 (indigo shade)
- FD&C Green No. 3 – Fast Green FCF, E143 (turquoise shade)
- FD&C Red No. 3 – Erythrosine, E127 (pink shade, commonly used in glacé cherries)
- FD&C Red No. 40 – Allura Red AC, E129 (red shade)
- FD&C Yellow No. 5 – Tartrazine, E102 (yellow shade)
- FD&C Yellow No. 6 – Sunset Yellow FCF, E110 (orange shade)

Permitted for Limited use in Foods

Two dyes are allowed by the FDA for limited applications:

- Citrus Red 2 (orange shade) - allowed only to color orange peels.
- Orange B (red shade) - allowed only for use in hot dog and sausage casings (not produced after 1978, but never delisted)

Delisted and Banned in The US

- FD&C Red No. 2 – Amaranth, E123
- FD&C Red No. 4
- FD&C Red No. 32 was used to color Florida oranges.
- FD&C Orange Number 1 was one of the first water-soluble dyes to be commercialized, and one of seven original food dyes allowed under the Pure Food and Drug Act of June 30, 1906.
- FD&C Orange No. 2 was used to color Florida oranges.
- FD&C Yellow No. 1, 2, 3, and 4
- FD&C Violet No. 1

Approved in EU

E numbers 102-143 cover the range of artificial colors. Some artificial dyes approved for food use in the EU include:

- E104: Quinoline Yellow
- E122: Carmoisine
- E124: Ponceau 4R
- E131: Patent Blue V
- E142: Green S

Natural Food Dyes

Natural food colors can make a variety of different hues

Carotenoids (E160, E161, E164), chlorophyllin (E140, E141), anthocyanins (E163), and betanin (E162) comprise four main categories of plant pigments grown to color food products. Other colorants or specialized derivatives of these core groups include:

- Annatto (E160b), a reddish-orange dye made from the seed of the achiote
- Caramel coloring (E150a-d), made from caramelized sugar
- Carmine (E120), a red dye derived from the cochineal insect, *Dactylopius coccus*
- Elderberry juice
- Lycopene (E160d)
- Paprika (E160c)
- Turmeric (E100)

Blue colors are especially rare.

To ensure reproducibility, the colored components of these substances are often provided in highly purified form. For stability and convenience, they can be formulated in suitable carrier materials (solid and liquids). Hexane, acetone, and other solvents break down cell walls in the fruit and vegetables and allow for maximum extraction of the coloring. Traces of these may still remain in the finished colorant, but they do not need to be declared on the product label. These solvents are known as carry-over ingredients.

Dyes and Lakes

Color additives are available for use in food as either "dyes" or lake pigments (commonly known as "lakes").

Dyes dissolve in water, but are not soluble in oil. Dyes are manufactured as powders, granules, liquids or other special purpose forms. They can be used in beverages, dry mixes, baked goods, confections, dairy products, pet foods, and a variety of other products. Dyes also have side effects that lakes lack. Consuming large amounts of dyes can color stools.

Lakes are made by combining dyes with salts (usually aluminum salts) to make insoluble compounds. Lakes tint by dispersion. Lakes are not oil-soluble, but are oil-dispersible. Lakes are more stable than dyes and are ideal for coloring products containing fats and oils or items lacking sufficient moisture to dissolve dyes. Typical uses include coated tablets, cake and doughnut mixes, hard candies and chewing gums, lipsticks, soaps, shampoos, talc, etc.

Other Uses

Because food dyes are generally safer to use than normal artists' dyes and pigments, some artists have used food coloring as a means of making pictures, especially in forms such as body-painting. Red food dye is often used in theatrical blood.

Most artificial food colorings are a type of acid dye, and can be used to dye protein fibers and nylon with the addition of an acid. They are all washfast and most are also lightfast. They will not permanently bond to plant fibers and other synthetics.

Criticism and Health Implications

There is no evidence to support broad claims that food coloring causes food intolerance and ADHD-like behavior in children. It is possible that certain food coloring may act as a trigger in those who are genetically predisposed, but the evidence is weak.

After concerns were expressed that food colorings may cause ADHD-like behavior in children, the collective evidence do not support this assertion. The US FDA and other food safety authorities to regularly review the scientific literature, and led the UK Food Standards Agency (FSA) to commission a study by researchers at Southampton University of the effect of a mixture of six food dyes (Tartrazine, Allura Red, Ponceau 4R, Quinoline Yellow WS, Sunset Yellow and Carmoisine (dubbed the "Southampton 6")) on children in the general population. These colorants are found in beverages. The study found "a possible link between the consumption of these artificial colours and a sodium benzoate preservative and increased hyperactivity" in the children; the advisory committee to the FSA that evaluated the study also determined that because of study limitations, the results could not be extrapolated to the general population, and further testing was recommended". The U.S. FDA did not make changes following the publication of the Southampton study, but following a citizen petition filed by the Center for Science in the Public Interest in 2008, requesting the FDA ban several food additives, the FDA reviewed the available evidence, and still made no changes.

The European regulatory community, with an emphasis on the precautionary principle, required labelling and temporarily reduced the acceptable daily intake (ADI) for the food colorings; the UK

FSA called for voluntary withdrawal of the colorings by food manufacturers. However, in 2009 the EFSA re-evaluated the data at hand and determined that "the available scientific evidence does not substantiate a link between the color additives and behavioral effects" for any of the dyes.

Flavor

Flavor, or flavour, is the sensory impression of food or other substance, and is determined primarily by the chemical senses of taste and smell. The "trigeminal senses", which detect chemical irritants in the mouth and throat as well as temperature and texture, are also important to the overall Gestalt of flavor perception. The flavor of the food, as such, can be altered with natural or artificial flavorants which affect these senses.

A *flavorant* is defined as a substance that gives another substance flavor, altering the characteristics of the solute, causing it to become sweet, sour, tangy, etc.

Of the three chemical senses, smell is the main determinant of a food item's flavor. While there are only five universally recognized basic tastes – sweet, sour, bitter, salty, and umami (savory) – the number of food smells is unbounded. A food's flavor, therefore, can be easily altered by changing its smell while keeping its taste similar. This is exemplified in artificially flavored jellies, soft drinks, and candies, which, while made of bases with a similar taste, have dramatically different flavors due to the use of different scents or fragrances. The flavorings of commercially produced food products are typically created by flavorists.

Although the terms *flavoring* and *flavorant* in common language denote the combined chemical sensations of taste and smell, the same terms are used in the fragrance and flavors industry to refer to edible chemicals and extracts that alter the flavor of food and food products through the sense of smell. Due to the high cost or unavailability of natural flavor extracts, most commercial flavorants are *nature-identical*, which means that they are the chemical equivalent of natural flavors but chemically synthesized rather than being extracted from source materials. Identification of nature-identical flavorants is done using technology such as headspace techniques.

Flavorants or Flavorings

Flavorings are focused on altering the flavors of natural food product such as meats and vegetables, or creating flavor for food products that do not have the desired flavors such as candies and other snacks. Most types of flavorings are focused on scent and taste. Few commercial products exist to stimulate the trigeminal senses, since these are sharp, astringent, and typically unpleasant flavors.

There are three principal types of flavorings used in foods, under definitions agreed in the EU and Australia:

Type	Description
Natural flavoring substances	Flavoring substances obtained from plant or animal raw materials, by physical, microbiological or enzymatic processes. They can be either used in their natural state or processed for human consumption, but cannot contain any nature-identical or artificial flavoring substances.

Nature-identical flavoring substances	Flavoring substances that are obtained by synthesis or isolated through chemical processes, which are chemically and organoleptically identical to flavoring substances naturally present in products intended for human consumption. They cannot contain any artificial flavoring substances.
Artificial flavoring substances	Flavoring substances not identified in a natural product intended for human consumption, whether or not the product is processed. These are typically produced by fractional distillation and additional chemical manipulation of naturally sourced chemicals, crude oil or coal tar. Although they are chemically different, in sensory characteristics are the same as natural ones.

Most artificial flavors are specific and often complex mixtures of singular naturally occurring flavor compounds combined together to either imitate or enhance a natural flavor. These mixtures are formulated by flavorists to give a food product a unique flavor and to maintain flavor consistency between different product batches or after recipe changes. The list of known flavoring agents includes thousands of molecular compounds, and the flavor chemist (flavorist) can often mix these together to produce many of the common flavors. Many flavorants consist of esters, which are often described as being "sweet" or "fruity".

Chemical	Odor
Diacetyl, Acetylpropionyl, Acetoin	Buttery
Isoamyl acetate	Banana
Benzaldehyde	Bitter almond, Cherry
Cinnamaldehyde	Cinnamon
Ethyl propionate	Fruity
Methyl anthranilate	Grape
Limonene	Orange
Ethyl decadienoate	Pear
Allyl hexanoate	Pineapple
Ethyl maltol	Sugar, Cotton candy
Ethylvanillin	Vanilla
Methyl salicylate	Wintergreen

The compounds used to produce artificial flavors are almost identical to those that occur naturally. It has been suggested that artificial flavors may be safer to consume than natural flavors due to the standards of purity and mixture consistency that are enforced either by the company or by law. Natural flavors in contrast may contain impurities from their sources while artificial flavors are typically more pure and are required to undergo more testing before being sold for consumption.

Flavors from food products are usually the result of a combination of natural flavors, which set up the basic smell profile of a food product while artificial flavors modify the smell to accent it.

Unlike smelling, which occurs upon inhalation, the sensing of flavors in the mouth occurs in the exhalation phase of breathing and is perceived differently by an individual. In other words, the smell of food is different depending on when you are smelling it in front of you or whether it has already entered your mouth.

Taste

While salt and sugar can technically be considered flavorants that enhance salty and sweet tastes, usually only compounds that enhance umami, as well as other secondary flavors are considered and referred to as *taste flavorants*. Artificial sweeteners are also technically flavorants.

Umami or "savory" flavorants, more commonly called taste or flavor enhancers, are largely based on amino acids and nucleotides. These are typically used as sodium or calcium salts. Umami flavorants recognized and approved by the European Union include:

Acid	Description
Glutamic acid salts	This amino acid's sodium salt, monosodium glutamate (MSG), is one of the most commonly used flavor enhancers in food processing. Mono and diglutamate salts are also commonly used.
Glycine salts	Simple amino acid salts typically combined with glutamic acid as flavor enhancers.
Guanylic acid salts	Nucleotide salts typically combined with glutamic acid as flavor enhancers.
Inosinic acid salts	Nucleotide salts created from the breakdown of AMP. Due to high costs of production, typically combined with glutamic acid as flavor enhancers.
5'-ribonucleotide salts	Nucleotide salts typically combined with other amino acids and nucleotide salts as flavor enhancers.

Certain organic and inorganic acids can be used to enhance sour tastes, but like salt and sugar these are usually not considered and regulated as flavorants under law. Each acid imparts a slightly different sour or tart taste that alters the flavor of a food.

Acid	Description
Acetic acid	Gives vinegar its sour taste and distinctive smell
Ascorbic acid	Found in oranges and green peppers and gives a crisp, slightly sour taste. Better known as vitamin C.
Citric acid	Found in citrus fruits and gives them their sour taste
Fumaric acid	Not found in fruits, used as a substitute for citric and tartaric acid
Lactic acid	Found in various milk or fermented products and give them a rich tartness
Malic acid	Found in apples and gives them their sour/tart taste
Phosphoric acid	Used in some cola drinks to give an acid taste
Tartaric acid	Found in grapes and wines and gives them a tart taste

Color

The color of food can affect one's expectations of the flavor significantly. In one study, adding more red color to a drink increased the perceived sweetness with darker colored solutions being rated 2–10% better than lighter ones, even though it had 1% less sucrose concentration.

Restrictions and Regulations

Regulations on Natural Flavoring

UK Food Law defines a natural flavor as:

a flavouring substance (or flavouring substances) which is (or are) obtained, by physical, enzymatic or microbiological processes, from material of vegetable or animal origin which material is either raw or has been subjected to a process normally used in preparing food for human consumption and to no process other than one normally so used.

The U.S. Code of Federal Regulations describes a "natural flavorant" as:

the essential oil, oleoresin, essence or extractive, protein hydrolysate, distillate, or any product of roasting, heating or enzymolysis, which contains the flavoring constituents derived from a spice, fruit or fruit juice, vegetable or vegetable juice, edible yeast, herb, bark, bud, root, leaf or any other edible portions of a plant, meat, seafood, poultry, eggs, dairy products, or fermentation products thereof, whose primary function in food is flavoring rather than nutritional.

The European Union's guidelines for natural flavorants are slightly different. Certain artificial flavorants are given an E number, which may be included on food labels.

Dietary Restrictions

Food manufacturers are sometimes reluctant to inform consumers about the source and identity of flavor ingredients and whether they have been produced with the incorporation of substances such as animal by-products. Some flavor ingredients, such as gelatin, are produced from animal products. Some, such as glycerin can be derived from either animal or vegetable sources. And some extracts, such as vanilla, may contain alcohol. Many Jews, Jains, Hindus, and Muslims adhere to religious dietary laws, and vegans to personal convictions, which restrict the use of animal by-products and/or alcohol in foods unless subject to oversight and inspection by their respective religious authority or moral beliefs. In many Western countries some consumers rely on a Jewish Kosher Pareve certification mark to indicate that natural flavorings used in a food product are free of meat and dairy (although they can still contain fish). The Vegan Society's Sunflower symbol (which is currently used by over 260 companies worldwide) can also be used to see which products do not use any animal ingredients (including flavorings and colorings).

Similarly, persons with known sensitivities or allergies to food products are advised to avoid foods that contain generic "natural flavors" or to first determine the source of the flavoring before consuming the food. Such flavors may be derived from a variety of source products that are themselves common allergens, such as dairy, soy, sesame, eggs, and nuts.

Flavor Creation

Food and beverage companies may require flavors for new products, product line extensions (e.g., low fat versions of existing products) or changes in formula or processing for existing products. In 2011, about US$10.6 billion were generated with the sale of flavors; the majority of the flavors used are consumed in processed and packaged food.

Most flavors represent a mixture of aroma compounds, the raw material that is produced by flavor companies. In rare cases, a single synthetic compound is used in pure form. Artificial vanilla flavors vanillin and ethylvanillin are a notable exception, as well as the artificial strawberry flavor (ethyl methylphenylglycidate). The ubiquitous "green apple" aroma is based on hexyl acetate.

The flavor creation is done by a specially trained scientist called a "flavorist". The flavorist's job combines scientific knowledge of the chemical palette with creativity to develop new and distinctive flavors. The flavor creation begins when the flavorist receives a brief from the client. In the brief the client will attempt to communicate exactly what type of flavor they seek, in what application it will be used, and any special requirements (e.g., must be all natural). The communication barrier can be quite difficult to overcome since most people aren't experienced at describing flavors. The flavorist will use his or her knowledge of the available chemical ingredients to create a formula and compound it on an electronic balance. The flavor will then be submitted to the client for testing. Several iterations, with feedback from the client, may be needed before the right flavor is found.

Additional work may also be done by the flavor company. For example, the flavor company may conduct sensory taste tests to test consumer acceptance of a flavor before it is sent to the client or to further investigate the "sensory space." The flavor company may also employ application specialists who work to ensure the flavor will work in the application for which it is intended. This may require special flavor delivery technologies that are used to protect the flavor during processing or cooking so that the flavor is only released when eaten by the end consumer.

Determination

Few standards are available or being prepared for sensory analysis of flavors. In chemical analysis of flavors, solid phase extraction (SPE), solid phase microextraction (SPME), and headspace gas chromatography are applied to extract and separate the flavor compounds in the sample. The determination is typically done by various mass spectrometric techniques.

Mineral (Nutrient)

A mineral is a chemical element required as an essential nutrient by organisms, other than carbon, hydrogen, nitrogen, oxygen and sulfur present in common organic molecules. These elements are classed as *minerals* in the four groups of essential nutrients; the others are vitamins, essential fatty acids, and essential amino acids.

Major chemical elements in order of abundance in the human body include calcium, phosphorus, potassium, sodium, chlorine, and magnesium. Important trace elements, necessary for mammalian life, include iron, cobalt, copper, zinc, manganese, molybdenum, iodine, and selenium. Because inorganic mineral content of foods do not form volatile combustion products, nutrition analysis methods involving combustion may report the total mineral content of food as "crude ash".

Over twenty minerals are necessary for mammals, and several more for various other types of life. The total number of chemical elements that are absolutely needed is not known for any organism.

Ultratrace elements of some minerals such as boron and chromium are known to clearly have a role but the exact biochemical nature is unknown, and others such as arsenic, bromine, and silicon are suspected to have a role in health, but with weaker evidence.

Most chemical elements that are ingested by organisms are in the form of simple compounds. Larger chemical compounds of elements need to be broken down for absorption. Plants absorb dissolved elements in soils, which are subsequently picked up by the herbivores that eat them, and the elements move up the food chain. Larger organisms may also consume soil (geophagia) and visit salt licks to obtain limiting minerals they are unable to acquire through other components of their diet.

Bacteria play an essential role in the weathering of primary elements that results in the release of nutrients for their own nutrition and for the nutrition of others in the ecological food chain. One element, cobalt, is available for use by animals only after having been processed into complex molecules (e.g., vitamin B12) by bacteria. Scientists are only recently starting to appreciate the magnitude and role that microorganisms have in the global cycling and formation of biominerals.

Essential Chemical Elements for Humans

At least twenty chemical elements are known to be *required* to support human biochemical processes by serving structural and functional roles as well as electrolytes. However, as many as twenty-nine elements in total (including the common hydrogen, carbon, nitrogen and oxygen) are suggested to be used by mammals, as a result of studies of biochemical, special uptake, and metabolic handling studies. However, many of these additional elements have no well-defined biochemical function known at present. Most of the known and suggested mineral nutrients are of relatively low atomic weight, and are reasonably common on land, or at least, common in the ocean (iodine, sodium):

Nutritional elements in the periodic table																		
H																		He
Li	Be												B	C	N	O	F	Ne
Na	Mg												Al	Si	P	S	Cl	Ar
K	Ca	Sc		Ti	V	Cr	Mn	Fe	Co	Ni	Cu	Zn	Ga	Ge	As	Se	Br	Kr
Rb	Sr	Y		Zr	Nb	Mo	Tc	Ru	Rh	Pd	Ag	Cd	In	Sn	Sb	Te	I	Xe
Cs	Ba	La	*	Hf	Ta	W	Re	Os	Ir	Pt	Au	Hg	Tl	Pb	Bi	Po	At	Rn
Fr	Ra	Ac	**	Rf	Db	Sg	Bh	Hs	Mt	Ds	Rg	Cn	Uut	Fl	Uup	Lv	Uus	Uuo
			*	Ce	Pr	Nd	Pm	Sm	Eu	Gd	Tb	Dy	Ho	Er	Tm	Yb	Lu	
			**	Th	Pa	U	Np	Pu	Am	Cm	Bk	Cf	Es	Fm	Md	No	Lr	

■ The four organic basic elements
■ Quantity elements
■ Essential trace elements
□ Suggested function from deprivation effects or active metabolic handling, but no clearly-identified biochemical function in humans
□ Limited circumstantial evidence for trace benefits or biological action in mammals

The following play important roles in biological processes:

DV = Daily Value (formerly Recommended Daily Allowance); UL = Tolerable Upper Intake Level (figures shown are for adults age 31-50, male or female neither pregnant nor lactating)

Dietary element	DV	UL	Amount	Category	High nutrient density dietary sources	Insufficiency	Excess
Potassium	3500	NE	Quantity	A systemic electrolyte and is essential in coregulating ATP with sodium	Sweetpotato, tomato, potato, beans, lentils, dairy products, seafood, banana, prune, carrot, orange	hypokalemia	hyperkalemia
Chlorine	3400	3600	Quantity	Needed for production of hydrochloric acid in the stomach and in cellular pump functions	Table salt (sodium chloride) is the main dietary source.	hypochloremia	hyperchloremia
Sodium	2400	2300	Quantity	A systemic electrolyte and is essential in coregulating ATP with potassium	Table salt (sodium chloride, the main source), sea vegetables, milk, and spinach.	hyponatremia	hypernatremia
Calcium	1000	2500	Quantity	Needed for muscle, heart and digestive system health, builds bone, supports synthesis and function of blood cells	Dairy products, eggs, canned fish with bones (salmon, sardines), green leafy vegetables, nuts, seeds, tofu, thyme, oregano, dill, cinnamon.	hypocalcaemia	hypercalcaemia

Dietary element	DV	UL	Amount	Category	High nutrient density dietary sources	Insufficiency	Excess
Phosphorus	1000	4000	Quantity	A component of bones, cells, in energy processing, in DNA and ATP (as phosphate) and many other functions	Red meat, dairy foods, fish, poultry, bread, rice, oats. In biological contexts, usually seen as phosphate	hypophosphatemia	hyperphosphatemia
Magnesium	400	350	Quantity	Required for processing ATP and for bones	Spinach, legumes, nuts, seeds, whole grains, peanut butter, avocado	hypomagnesemia, magnesium deficiency	hypermagnesemia
Iron	18	45	Trace	Required for many proteins and enzymes, notably hemoglobin to prevent anemia	Meat, seafood, nuts, beans, dark chocolate	iron deficiency	iron overload disorder
Zinc	15	40	Trace	Pervasive and required for several enzymes such as carboxypeptidase, liver alcohol dehydrogenase, and carbonic anhydrase	Oysters*, red meat, poultry, nuts, whole grains, dairy products	zinc deficiency	zinc toxicity
Manganese	2	350	Trace	A cofactor in enzyme functions	Grains, legumes, seeds, nuts, leafy vegetables, tea, coffee	manganese deficiency	manganism
Copper Main article: Copper in health	2	11	Trace	Required component of many redox enzymes, including cytochrome c oxidase	Liver, seafood, oysters, nuts, seeds; some: whole grains, legumes	copper deficiency	copper toxicity
Iodine	0.150	1.1	Trace	Required for synthesis of thyroid hormones, thyroxine and triiodothyronine and to prevent goiter: • Iodine in biology	Seaweed (kelp or kombu)*, grains, eggs, iodized salt	iodine deficiency	iodism Hyperthyroidism

Dietary element	DV	UL	Amount	Category	High nutrient density dietary sources	Insufficiency	Excess
Chromium	0.120	NE	Trace	Involved in glucose and lipid metabolism, although its mechanisms of action in the body and the amounts needed for optimal health are not well defined	Broccoli, grape juice (especially red), meat, whole grain products	Chromium deficiency	Chromium toxicity
Molybdenum	0.075	2	Trace	The oxidases xanthine oxidase, aldehyde oxidase, and sulfite oxidase	Legumes, whole grains, nuts	molybdenum deficiency	molybdenum toxicity
Selenium	0.070	0.4	Trace	Essential to activity of antioxidant enzymes like glutathione peroxidase	Brazil nuts, seafoods, organ meats, meats, grains, dairy products, eggs	selenium deficiency	selenosis
Cobalt	none	NE	Trace	Required in the synthesis of vitamin B_{12}, but because bacteria are required to synthesize the vitamin, it is usually considered part of vitamin B_{12}			Cobalt poisoning

* One serving exceeds UL (tolerable upper intake level).

Blood Concentrations of Minerals

Minerals are present in a healthy human being's blood at certain mass and molar concentrations. The figure below presents the concentrations of each of the chemical elements discussed in this article, from center-right to the right. Depending on the concentrations, some are in upper part of the picture, while others are in the lower part. The figure includes the relative values of other constituents of blood such as hormones. In the figure, minerals are color highlighted in purple.

Dietary Nutrition

Dietitians may recommend that minerals are best supplied by ingesting specific foods rich with the chemical element(s) of interest. The elements may be naturally present in the food (e.g., calcium in dairy milk) or added to the food (e.g., orange juice fortified with calcium; iodized salt, salt fortified with iodine). Dietary supplements can be formulated to contain several different chemical elements (as compounds), a combination of vitamins and/or other chemical compounds, or a single element (as a compound or mixture of compounds), such as calcium (as calcium carbonate, calcium citrate, etc.) or magnesium (as magnesium oxide, etc.), chromium (usually as chromium(III) picolinate), or iron (as iron bis-glycinate).

The dietary focus on chemical elements derives from an interest in supporting the biochemical reactions of metabolism with the required elemental components. Appropriate intake levels of certain chemical elements have been demonstrated to be required to maintain optimal health. Diet can meet all the body's chemical element requirements, although supplements can be used when some requirements (e.g., calcium, which is found mainly in dairy products) are not adequately met by the diet, or when chronic or acute deficiencies arise from pathology, injury, etc. Research has supported that altering inorganic mineral compounds (carbonates, oxides, etc.) by reacting them with organic ligands (amino acids, organic acids, etc.) improves the bioavailability of the supplemented mineral.

Elements Considered Possibly Essential but not Confirmed

Many ultratrace elements have been suggested as essential, but such claims have usually not been confirmed. Definitive evidence for efficacy comes from the characterization of a biomolecule containing the element with an identifiable and testable function. One problem with identifying efficacy is that some elements are innocuous at low concentrations and are pervasive (examples: silicon and nickel in solid and dust), so proof of efficacy is lacking because deficiencies are difficult to reproduce.

Element	Description	Excess
Bromine	Possibly important to basement membrane architecture and tissue development.	bromism
Arsenic	Essential in rat, hamster, goat and chicken models, but no biochemical mechanism known in humans.	arsenic poisoning
Nickel	Nickel is an essential component of several enzymes, including urease and hydrogenase. Although not required by humans, some are thought to be required by gut bacteria, such as urease required by some varieties of Bifidobacterium. In humans, nickel may be a cofactor or structural component of certain metalloenzymes involved in hydrolysis, redox rections, and gene expression. Nickel deficiency depressed growth in goats, pigs, and sheep, and diminished circulating thyroid hormone concentration in rats.	Nickel toxicity
Fluorine	Fluorine (as fluoride) is not generally considered an essential element because humans do not require it for growth or to sustain life. However, if one considers the prevention of dental cavities an important criterion in determining essentiality, then fluoride might well be considered an essential trace element. However, recent research indicates that the primary action of fluoride occurs topically (at the surface).	Fluoride poisoning
Boron	Boron is an essential plant nutrient, required primarily for maintaining the integrity of cell walls. Boron has been shown to be essential to complete the life cycle in representatives of all phylogenetic kingdoms, including the model species danio rerio (zebrafish) and Xenopus laevis (African clawed frog). In animals, supplemental boron has been shown to reduce calcium excretion and activate vitamin D.	Nontoxic
Lithium	It is not known whether lithium has a physiological role in any species, but nutritional studies in mammals have indicated its importance to health, leading to a suggestion that it be classed as an essential trace element.	Lithium toxicity

Strontium	Strontium has been found to be involved in the utilization of calcium in the body. It has promoting action on calcium uptake into bone at moderate dietary strontium levels, but a rachitogenic (rickets-producing) action at higher dietary levels.	Rachitogenic (causing Rickets)
Other	Silicon and vanadium have established, albeit specialized, biochemical roles as structural or functional cofactors in other organisms, and are possibly, even probably, used by mammals (including humans). By contrast, tungsten and cadmium have specialized biochemical uses in certain lower organisms, but these elements appear not to be utilized by humans. Other elements considered to be possibly essential include aluminium, germanium, lead, rubidium, and tin.	Multiple

Mineral Ecology

Recent studies have shown a tight linkage between living organisms and chemical elements on this planet. This led to the redefinition of minerals as "an element or compound, amorphous or crystalline, formed through 'biogeochemical' processes. The addition of `bio' reflects a greater appreciation, although an incomplete understanding, of the processes of mineral formation by living forms." Biologists and geologists have only recently started to appreciate the magnitude of mineral biogeoengineering. Bacteria have contributed to the formation of minerals for billions of years and critically define the biogeochemical mineral cycles on this planet. Microorganisms can precipitate metals from solution contributing to the formation of ore deposits in addition to their ability to catalyze mineral dissolution, to respire, precipitate, and form minerals.

Most minerals are inorganic in nature. Mineral nutrients refers to the smaller class of minerals that are metabolized for growth, development, and vitality of living organisms. Mineral nutrients are recycled by bacteria that are freely suspended in the vast water columns of the worlds oceans. They absorb dissolved organic matter containing mineral nutrients as they scavenge through the dying individuals that fall out of large phytoplankton blooms. Flagellates are effective bacteriovores and are also commonly found in the marine water column. The flagellates are preyed upon by zooplankton while the phytoplankton concentrates on the larger particulate matter that is suspended in the water column as they are consumed by larger zooplankton, with fish as the top predator. Mineral nutrients cycle through this marine food chain, from bacteria and phytoplankton to flagellates and zooplankton who are then eaten by fish. The bacteria are important in this chain because only they have the physiological ability to absorb the dissolved mineral nutrients from the sea. These recycling principals from marine environments apply to many soil and freshwater ecosystems as well.

References

- Bucci, Luke (1995). Nutrition applied to injury rehabilitation and sports medicine. Boca Raton: CRC Press. p. 151. ISBN 0-8493-7913-X.
- Assessment of technologies for determining cancer risks from the environment. Darby, Pennsylvania, USA: DIANE publishing. 1981. p. 177. ISBN 1-4289-2437-X.
- Tomaska LD and Brooke-Taylor, S. Food Additives - General pp 449-454 in Encyclopedia of Food Safety, Vol 2: Hazards and Diseases. Eds, Motarjemi Y et al. Academic Press, 2013. ISBN 9780123786135
- Corbridge, D. E. C. (1995-02-01). Phosphorus: An Outline of Its Chemistry, Biochemistry, and Technology (5th ed.). Amsterdam: Elsevier Science Pub Co. p. 1220. ISBN 0-444-89307-5.
- Schlenker, Eleanor; Gilbert, Joyce Ann (28 August 2014). Williams' Essentials of Nutrition and Diet Therapy.

Elsevier Health Sciences. pp. 162–3. ISBN 978-0-323-29401-0. Retrieved 15 July 2016.

- Jameson, J. Larry; De Groot, Leslie J. (25 February 2015). Endocrinology: Adult and Pediatric. Elsevier Health Sciences. p. 1510. ISBN 978-0-323-32195-2. Retrieved 14 July 2016.
- Kim, Myoung Jin; Anderson, John; Mallory, Caroline (1 February 2014). Human Nutrition. Jones & Bartlett Publishers. p. 241. ISBN 978-1-4496-4742-1. Retrieved 10 July 2016.
- Gropper, Sareen S.; Smith, Jack L. (1 June 2012). Advanced Nutrition and Human Metabolism. Cengage Learning. pp. 527–8. ISBN 1-133-10405-3. Retrieved 10 July 2016.
- Lippard, Stephen J.; Jeremy M. Berg (1994). Principles of Bioinorganic Chemistry. Mill Valley, CA: University Science Books. p. 411. ISBN 0-935702-72-5.
- Berdanier, Carolyn D.; Dwyer, Johanna T.; Heber, David (19 April 2016). Handbook of Nutrition and Food, Third Edition. CRC Press. pp. 211–26. ISBN 978-1-4665-0572-8. Retrieved 3 July 2016.
- Sigel, Astrid; Sigel, Helmut; Sigel, Roland K. O. (27 January 2014). Interrelations between Essential Metal Ions and Human Diseases. Springer Science & Business Media. p. 349. ISBN 978-94-007-7500-8. Retrieved 4 July 2016.
- Institute of Medicine (29 September 2006). Dietary Reference Intakes: The Essential Guide to Nutrient Requirements. National Academies Press. pp. 313–19, 415–22. ISBN 978-0-309-15742-1.
- Erdman, John W., Jr.; MacDonald, Ian A.; Zeisel, Steven H. (30 May 2012). Present Knowledge in Nutrition. John Wiley & Sons. p. 1324. ISBN 978-0-470-96310-4. Retrieved 4 July 2016.
- Gottschlich, Michele M. (2001). The Science and Practice of Nutrition Support: A Case-based Core Curriculum. Kendall Hunt. p. 98. ISBN 978-0-7872-7680-5. Retrieved 9 July 2016.
- Arlt, Ulrike (29 Apr 2011). "The Legislation of Food Colours in Europe". The Natural Food Colours Association. Retrieved 18 Feb 2014.
- Downham, Alison; Collins, Paul (2000). "Colouring our foods in the last and next millennium" (PDF). International Journal of Food Science and Technology. Blackwell Science Ltd. 35: 5–22. doi:10.1046/j.1365-2621.2000.00373.x. Retrieved 18 Feb 2014.
- Meadows, Michelle (2006). "A Century of Ensuring Safe Foods and Cosmetics". FDA Consumer magazine. FDA (January–February). Retrieved 21 Feb 2014.

5

Food Adulteration: Current Issues and Challenges

Adulterants in food are unwanted substances that reduce the quality of food or make it harmful. They contaminate the food and make it unfit for consumption. Some adulterants are added on purpose to act as fillers in expensive food products. This chapter discusses the various types of adulterants such as aldicarb, lead and melamine.

Adulterant

An adulterant is a pejorative term for a substance found within other substances such as food, fuels or chemicals, although not allowed for legal or other reasons. It will not normally be present in any specification or declared contents of the substance, and may not be legally allowed. The addition of adulterants is called adulteration. The most common reason for adulteration is the use by manufacturers of undeclared materials that are cheaper than the correct and declared ones. The adulterants may be harmful, or reduce the potency of the product, or they may be harmless.

The term "contamination" is usually used for the inclusion of unwanted substances due to accident or negligence rather than intent, and also for the introduction of unwanted substances after the product has been made. Adulteration therefore implies that the adulterant was introduced deliberately in the initial manufacturing process, or sometimes that it was present in the raw materials and should have been removed, but was not.

An adulterant is distinct from, for example, permitted food additives. There can be a fine line between adulterant and additive; chicory may be added to coffee to reduce the cost or achieve a desired flavour—this is adulteration if not declared, but may be stated on the label. Chalk was often added to bread flour; this reduces the cost and increases whiteness, but the calcium actually confers health benefits, and in modern bread a little chalk may be included as an additive for this reason.

In wartime adulterants have been added to make foodstuffs "go further" and prevent shortages. The German word ersatz is widely recognised from such practices during WW2. Such adulteration was sometimes deliberately hidden from the population to prevent loss of morale and propaganda reasons. Some goods considered luxurious in the Communist Bloc such as coffee were adulterated to make them affordable to the general population.

Adulterants added to reduce the amount of expensive product in illicit drugs are called cutting agents. Deliberate addition of toxic adulterants to food or other products for human consumption is poisoning.

In Food and Beverages

Past and present examples of adulteration, some dangerous, include:

- Roasted chicory roots used as an adulterant for coffee
- Diethylene glycol, used dangerously by some winemakers in sweet wines
- Apple jellies (jams), as substitutes for more expensive fruit jellies, with added colorant and sometimes even specks of wood that simulate raspberry or strawberry seeds
- Water, for diluting milk and alcoholic beverages
- Cutting agents used to adulterate (or "cut") illicit drugs—for example, shoe polish in hashish, amphetamines in ecstasy, lactose in cocaine
- Urea, melamine and other nonprotein nitrogen sources, added to protein products to inflate crude protein content measurements
- High fructose corn syrup or cane sugar, used to adulterate honey
- Water or brine injected into chicken, pork, or other meats to increase their weight

History

Historically, the use of adulterants has been common; sometimes dangerous substances have been used. In the United Kingdom up to the Victorian era, adulterants were common; for example, cheeses were sometimes colored with lead. Similar adulteration issues were seen in industry in the United States, during the 19th century. There is dispute over whether these practices declined primarily due to government regulation or to increased public awareness and concern over the practices.

In the early 21st century, cases of dangerous adulteration occurred in the People's Republic of China.

In some African countries, it is not uncommon for thieves to break electric transformers to steal transformer oil, which is then sold to the operators of roadside food stalls to be used for deep frying. When used for frying, it is reported that transformer oil lasts much longer than regular cooking oil. The downside of this misuse of the transformer oil is the threat to the health of the consumers, due to the presence of PCBs.

Adulterant use was first investigated in 1820 by the German chemist Frederick Accum, who identified many toxic metal colorings in food and drink. His work antagonized food suppliers, and he was ultimately discredited by a scandal over his alleged mutilation of books of the Royal Institution library. The physician Arthur Hill Hassall conducted extensive studies in the early 1850s, which were published in *The Lancet* and led to the 1860 Food Adulteration Act and other legislation. John Postgate led a further campaign, leading to another Act of 1875, which forms the basis of the modern legislation and a system of public analysts who test for adulteration.

At the turn of the 20th century, industrialization in the United States led to a rise in adulteration which inspired some protest. Accounts of adulteration led the New York *Evening Post* to parody:

Mary had a little lamb,

And when she saw it sicken,

She shipped it off to Packingtown,

And now it's labeled chicken.

However, even in the 18th century, people complained about adulteration in food:

"The bread I eat in London is a deleterious paste, mixed up with chalk, alum and bone ashes, insipid to the taste and destructive to the constitution. The good people are not ignorant of this adulteration; but they prefer it to wholesome bread, because it is whiter than the meal of corn [wheat]. Thus they sacrifice their taste and their health. . . to a most absurd gratification of a misjudged eye; and the miller or the baker is obliged to poison them and their families, in order to live by his profession." – Tobias Smollet, *The Expedition of Humphrey Clinker* (1771)

Incidents of Adulteration

- In 1987, Beech-Nut was fined for violating the US Federal Food, Drug, and Cosmetic Act by selling flavored sugar water as apple juice.
- In 1997, ConAgra Foods illegally sprayed water on stored grain to increase its weight.
- In 2007, samples of wheat gluten mixed with melamine, presumably to produce inflated results from tests for protein content, were discovered in the USA. They were found to have come from China.
- In 2008, significant portions of China's milk supply were found to have been adulterated with melamine. Infant formula produced from this milk killed at least six children and is believed to have harmed thousands of others.
- In 2012, a study in India across 29 states and union territories found that milk was adulterated with detergent, fat, and even urea, and diluted with water. Just 31.5% of samples conformed to FSSAI standards.
- In the 2013 meat adulteration scandal in Europe, horsemeat was passed off as beef.

Various Types of Adulterants:

Aldicarb

Aldicarb is a carbamate insecticide which is the active substance in the pesticide Temik. It is effective against thrips, aphids, spider mites, lygus, fleahoppers, and leafminers, but is primarily used as a nematicide. Aldicarb is a cholinesterase inhibitor which prevents the breakdown of acetylcholine in the synapse. In case of severe poisoning, the victim dies of respiratory failure.

Aldicarb is one of the most widely used pesticides internationally, and is also one of the most environmentally toxic. Aldicarb poisoning from agricultural water runoff has led to the destruction of healthy ecosystems and the irreversible poisoning of fertile agricultural land. Poisoning from this pesticide is also believed to be linked to high cancer rates in communities located around the Aral Sea.

Aldicarb is effective where resistance to organophosphate insecticides has developed, and is extremely important in potato production, where it is used for the control of soil-borne nematodes and some foliar pests. Its high level of solubility restricts its use in certain areas where the water table is close to the surface.

Regulatory Status

Aldicarb was approved by the USEPA for use by professional pesticide applicators on a variety of crops, including cotton, beans, and others. It is not approved for household use.

Aldicarb was one of the "dirty dozen" pesticides that the environmental group Pesticide Action Network North America targeted in 1985. EPA put a ban on the main aldicarb pesticide, Temik 15G (Produced by Bayer CropScience), in 2010, requiring an end to distribution by 2017. Use on citrus and potatoes was banned beginning in 2012, with a complete phase out of the product expected by 2018. Despite this, a new aldicarb pesticide named Meymik 15G, produced by AgLogic Chemical, was approved by the EPA in December 2011 and is said to be entering the market in 2015. It will be registered for use on cotton, dry beans, peanuts, soybeans, sugar beets, and sweet potatoes.

"Tres Pasitos", a mouse, rat, and roach killer that contains high concentrations of aldicarb, has been illegally imported into the United States from Mexico and other Latin American countries. The product is highly toxic to animals and people, and according to the EPA "should never be used in [the] home."

It is classified as an extremely hazardous substance in the United States as defined in Section 302 of the U.S. Emergency Planning and Community Right-to-Know Act (42 U.S.C. 11002), and is subject to strict reporting requirements by facilities which produce, store, or use it in significant quantities.

History

Aldicarb is manufactured by Bayer CropScience, but was formerly owned and produced by Union Carbide. Union Carbide's agricultural chemicals division was sold to Rhône-Poulenc. Later, Aventis Cropscience was formed from Hoechst AG and Rhone-Poulenc Agrochemical, which lasted until Bayer acquired it in 2002.

In November 2009, corn treated with Temik was placed in and around peanut fields in Eastland County, Texas, near the town of Cisco. The corn was eaten by feral hogs, deer, and other animals, prompting the Texas Parks and Wildlife Department to issue a hunting ban.

Toxicity in Mammals

Aldicarb is a fast-acting cholinesterase inhibitor, causing rapid accumulation of acetylcholine at the synaptic cleft. It is widely used to study cholinergic neurotransmission in simple systems such as the nematode *C. elegans*.

Exposure to high amounts of aldicarb can cause weakness, blurred vision, headache, nausea, tearing, sweating, and tremors in humans. High doses can be fatal to humans because it can paralyze the respiratory system.

In South Africa (where Aldicarb is popularly known as Two Step) it is widely used by burglars to poison dogs.

Lead Poisoning

Lead poisoning is a type of metal poisoning caused by increased levels of the heavy metal lead in the body. Like most toxic heavy metals, lead interferes with a variety of body processes and is toxic to many organs and tissues, including the heart, bones, intestines, kidneys, and reproductive and nervous systems. The brain is the organ most sensitive to lead exposure. Lead interferes with the development of the nervous system and is therefore particularly toxic to children, causing potentially permanent learning and behavior disorders including violence. Symptoms include abdominal pain, confusion, headache, anemia, irritability, and in severe cases seizures, coma, and death.

Routes of exposure to lead include contaminated air, water, soil, food, and consumer products. Occupational exposure is a common cause of lead poisoning in adults. According to estimates made by the National Institute of Occupational Safety and Health (NIOSH), more than 3 million workers in the United States are potentially exposed to lead in the workplace. One of the largest threats to children is lead paint that exists in many homes, especially older ones; thus children in older housing with chipping paint or lead dust from moveable window frames with lead paint are at greater risk. Prevention of lead exposure can range from individual efforts (e.g., removing lead-containing items such as piping or blinds from the home) to nationwide policies (e.g., laws that ban lead in products, reduce allowable levels in water or soil, or provide for cleanup and mitigation of contaminated soil, etc.)

Elevated lead in the body can be detected by the presence of changes in blood cells visible with a microscope and dense lines in the bones of children seen on X-ray, but the main tool for diagnosis is measurement of the blood lead level. When blood lead levels are recorded, the results indicate how much lead is circulating within the blood stream, not the amount stored in the body. There are two units for reporting blood lead level, either micrograms per deciliter (µg/dl), or micrograms per 100 grams (µg/100 g) of whole blood, which are numerically equivalent. The Centers for Disease Control (US) has set the standard elevated blood lead level for adults to be 10 µg/dl of the whole blood. For children the number is set much lower at 5 µg/dl of blood as of 2012 down from a previous 10 µg/dl. Children are especially prone to the health effects of lead. As a result, blood lead levels must be set lower and closely monitored if contamination is possible. The major treatments are removal of the source of lead and chelation therapy (administration of agents that bind lead so it can be excreted).

Humans have been mining and using this heavy metal for thousands of years, poisoning themselves in the process. Although lead poisoning is one of the oldest known work and environmental hazards, the modern understanding of the small amount of lead necessary to cause harm did not come about until the latter half of the 20th century. No safe threshold for lead exposure has been discovered—that is, there is no known sufficiently small amount of lead that will not cause harm to the body.

Classification

Classically, "lead poisoning" or "lead intoxication" has been defined as exposure to high levels of lead typically associated with severe health effects. Poisoning is a pattern of symptoms that occur with toxic effects from mid to high levels of exposure; toxicity is a wider spectrum of effects, including subclinical ones (those that do not cause symptoms). However, professionals often use "lead poisoning" and "lead toxicity" interchangeably, and official sources do not always restrict the use of "lead poisoning" to refer only to symptomatic effects of lead.

The amount of lead in the blood and tissues, as well as the time course of exposure, determine toxicity. Lead poisoning may be acute (from intense exposure of short duration) or chronic (from repeat low-level exposure over a prolonged period), but the latter is much more common. Diagnosis and treatment of lead exposure are based on blood lead level (the amount of lead in the blood), measured in micrograms of lead per deciliter of blood (μg/dL). Urine lead levels may be used as well, though less commonly. In cases of chronic exposure lead often sequesters in the highest concentrations first in the bones, then in the kidneys. If a provider is performing a provocative excretion test, or "chelation challenge", a measurement obtained from urine rather than blood is likely to provide a more accurate representation of total lead burden to a skilled interpreter.

The US Centers for Disease Control and Prevention and the World Health Organization state that a blood lead level of 10 μg/dL or above is a cause for concern; however, lead may impair development and have harmful health effects even at lower levels, and there is no known safe exposure level. Authorities such as the American Academy of Pediatrics define lead poisoning as blood lead levels higher than 10 μg/dL.

Lead forms a variety of compounds and exists in the environment in various forms. Features of poisoning differ depending on whether the agent is an organic compound (one that contains carbon), or an inorganic one. Organic lead poisoning is now very rare, because countries across the world have phased out the use of organic lead compounds as gasoline additives, but such compounds are still used in industrial settings. Organic lead compounds, which cross the skin and respiratory tract easily, affect the central nervous system predominantly.

Signs and Symptoms

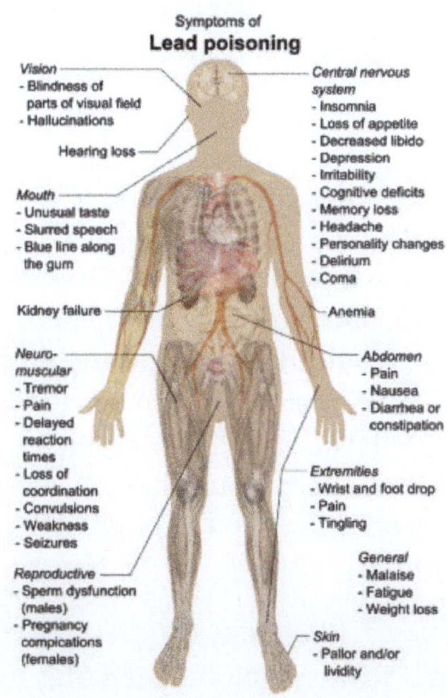

Symptoms of lead poisoning.

Lead poisoning can cause a variety of symptoms and signs which vary depending on the individual and the duration of lead exposure. Symptoms are nonspecific and may be subtle, and someone with elevated lead levels may have no symptoms. Symptoms usually develop over weeks to months as lead builds up in the body during a chronic exposure, but acute symptoms from brief, intense exposures also occur. Symptoms from exposure to organic lead, which is probably more toxic than inorganic lead due to its lipid solubility, occur rapidly. Poisoning by organic lead compounds has symptoms predominantly in the central nervous system, such as insomnia, delirium, cognitive deficits, tremor, hallucinations, and convulsions.

Symptoms may be different in adults and children; the main symptoms in adults are headache, abdominal pain, memory loss, kidney failure, male reproductive problems, and weakness, pain, or tingling in the extremities.

Early symptoms of lead poisoning in adults are commonly nonspecific and include depression, loss of appetite, intermittent abdominal pain, nausea, diarrhea, constipation, and muscle pain. Other early signs in adults include malaise, fatigue, decreased libido, and problems with sleep. An unusual taste in the mouth and personality changes are also early signs.

In adults, symptoms can occur at levels above 40 μg/dL, but are more likely to occur only above 50–60 μg/dL. Symptoms begin to appear in children generally at around 60 μg/dL. However, the lead levels at which symptoms appear vary widely depending on unknown characteristics of each individual. At blood lead levels between 25 and 60 μg/dL, neuropsychiatric effects such as delayed reaction times, irritability, and difficulty concentrating, as well as slowed motor nerve conduction and headache can occur. Anemia may appear at blood lead levels higher than 50 μg/dL. In adults, abdominal colic, involving paroxysms of pain, may appear at blood lead levels greater than 80 μg/dL. Signs that occur in adults at blood lead levels exceeding 100 μg/dL include wrist drop and foot drop, and signs of encephalopathy (a condition characterized by brain swelling), such as those that accompany increased pressure within the skull, delirium, coma, seizures, and headache. In children, signs of encephalopathy such as bizarre behavior, discoordination, and apathy occur at lead levels exceeding 70 μg/dL. For both adults and children, it is rare to be asymptomatic if blood lead levels exceed 100 μg/dL.

Acute Poisoning

In acute poisoning, typical neurological signs are pain, muscle weakness, numbness and tingling, and, rarely, symptoms associated with inflammation of the brain. Abdominal pain, nausea, vomiting, diarrhea, and constipation are other acute symptoms. Lead's effects on the mouth include astringency and a metallic taste. Gastrointestinal problems, such as constipation, diarrhea, poor appetite, or weight loss, are common in acute poisoning. Absorption of large amounts of lead over a short time can cause shock (insufficient fluid in the circulatory system) due to loss of water from the gastrointestinal tract. Hemolysis (the rupture of red blood cells) due to acute poisoning can cause anemia and hemoglobin in the urine. Damage to kidneys can cause changes in urination such as decreased urine output. People who survive acute poisoning often go on to display symptoms of chronic poisoning.

Chronic Poisoning

Chronic poisoning usually presents with symptoms affecting multiple systems, but is associated with three main types of symptoms: gastrointestinal, neuromuscular, and neurological. Central nervous system and neuromuscular symptoms usually result from intense exposure, while gastrointestinal symptoms usually result from exposure over longer periods. Signs of chronic exposure include loss of short-term memory or concentration, depression, nausea, abdominal pain, loss of coordination, and numbness and tingling in the extremities. Fatigue, problems with sleep, headaches, stupor, slurred speech, and anemia are also found in chronic lead poisoning. A "lead hue" of the skin with pallor and/or lividity is another feature. A blue line along the gum with bluish black edging to the teeth, known as a Burton line, is another indication of chronic lead poisoning. Children with chronic poisoning may refuse to play or may have hyperkinetic or aggressive behavior disorders. Visual disturbance may present with gradually progressing blurred vision as a result of central scotoma, caused by toxic optic neuritis.

Effects on Children

A fetus developing in the womb of a woman who has elevated blood lead level is susceptible to lead poisoning by intrauterine exposure, and is at greater risk of being born prematurely or with a low birth weight.

Children are more at risk for lead poisoning because their smaller bodies are in a continuous state of growth and development. Lead is absorbed at a faster rate compared to adults, which causes more physical harm than to older people. Furthermore, children, especially as they are learning to crawl and walk, are constantly on the floor and therefore more prone to ingesting and inhaling dust that is contaminated with lead.

The classic signs and symptoms in children are loss of appetite, abdominal pain, vomiting, weight loss, constipation, anemia, kidney failure, irritability, lethargy, learning disabilities, and behavioral problems. Slow development of normal childhood behaviors, such as talking and use of words, and permanent intellectual disability are both commonly seen. Although less common, it is possible for fingernails to develop leukonychia striata if exposed to abnormally high lead concentrations.

Complications

Lead affects every one of the body's organ systems, especially the nervous system, but also the bones and teeth, the kidneys, and the cardiovascular, immune, and reproductive systems. Hearing loss and tooth decay have been linked to lead exposure, as have cataracts. Intrauterine and neonatal lead exposure promote tooth decay. Aside from the developmental effects unique to young children, the health effects experienced by adults are similar to those in children, although the thresholds are generally higher.

Kidneys

Kidney damage occurs with exposure to high levels of lead, and evidence suggests that lower levels can damage kidneys as well. The toxic effect of lead causes nephropathy and may cause Fanconi syndrome, in which the proximal tubular function of the kidney is impaired. Long-term exposure at levels lower than those that cause lead nephropathy have also been reported as nephrotoxic in patients from developed countries that had chronic kidney disease or were at risk because of hypertension or diabetes mellitus. Lead poisoning inhibits excretion of the waste product urate and causes a predisposition for gout, in which urate builds up. This condition is known as *saturnine gout*.

Cardiovascular System

Evidence suggests lead exposure is associated with high blood pressure, and studies have also found connections between lead exposure and coronary heart disease, heart rate variability, and death from stroke, but this evidence is more limited. People who have been exposed to higher concentrations of lead may be at a higher risk for cardiac autonomic dysfunction on days when ozone and fine particles are higher.

Reproductive System

Lead affects both the male and female reproductive systems. In men, when blood lead levels exceed 40 μg/dL, sperm count is reduced and changes occur in volume of sperm, their motility, and their morphology. A pregnant woman's elevated blood lead level can lead to miscarriage, prematurity, low birth weight, and problems with development during childhood. Lead is able to pass through the placenta and into breast milk, and blood lead levels in mothers and infants are usually similar. A fetus may be poisoned *in utero* if lead from the mother's bones is subsequently mobilized by the changes in metabolism due to pregnancy; increased calcium intake in pregnancy may help mitigate this phenomenon.

Nervous System

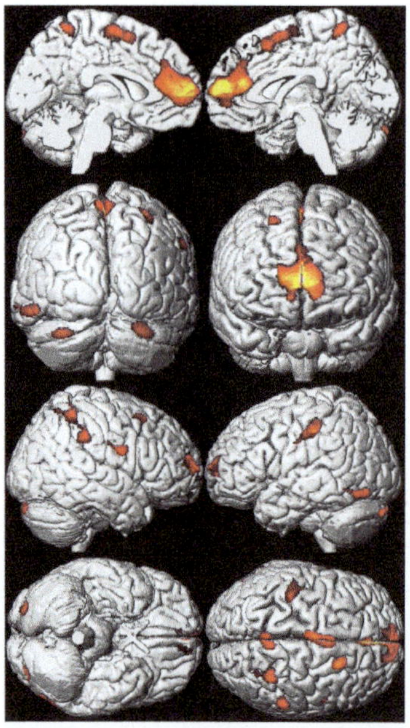

The brains of adults who were exposed to lead as children show decreased volume, especially in the prefrontal cortex, on MRI. Areas of volume loss are shown in color over a template of a normal brain.

Lead affects the peripheral nervous system (especially motor nerves) and the central nervous system. Peripheral nervous system effects are more prominent in adults and central nervous system effects are more prominent in children. Lead causes the axons of nerve cells to degenerate and lose their myelin coats.

Lead exposure in young children has been linked to learning disabilities, and children with blood lead concentrations greater than 10 μg/dL are in danger of developmental disabilities. Increased blood lead level in children has been correlated with decreases in intelligence, nonverbal reasoning, short-term memory, attention, reading and arithmetic ability, fine motor skills, emotional regulation, and social engagement. The effect of lead on children's cognitive abilities takes place at very low levels. There is apparently no lower threshold to the dose-response relationship (unlike other heavy metals such as mercury). Reduced academic performance has been associated with lead exposure even at blood lead levels lower than 5 μg/dL. Blood lead levels below 10 μg/dL have been reported to be associated with lower IQ and behavior problems such as aggression, in proportion with blood lead levels. Between the blood lead levels of 5 and 35 μg/dL, an IQ decrease of 2–4 points for each μg/dL increase is reported in children.

High blood lead levels in adults are also associated with decreases in cognitive performance and with psychiatric symptoms such as depression and anxiety. It was found in a large group of current and former inorganic lead workers in Korea that blood lead levels in the range of 20–50 μg/dL were correlated with neuro-cognitive defects. Increases in blood lead levels from about 50 to about 100 μg/dL in adults have been found to be associated with persistent, and possibly permanent, impairment of central nervous system function.

Lead exposure in children is also correlated with neuropsychiatric disorders such as attention deficit hyperactivity disorder and anti-social behaviour. Elevated lead levels in children are correlated with higher scores on aggression and delinquency measures. A correlation has also been found between prenatal and early childhood lead exposure and violent crime in adulthood. Countries with the highest air lead levels have also been found to have the highest murder rates, after adjusting for confounding factors. A May 2000 study by economic consultant Rick Nevin theorizes that lead exposure explains 65% to 90% of the variation in violent crime rates in the US. A 2007 paper by the same author claims to show a strong association between preschool blood lead and subsequent crime rate trends over several decades across nine countries. It is believed that the U.S. ban on lead paint in buildings in the late 1970s, as well as the phaseout of leaded gasoline in the 1970s and 1980s, partially helped contribute to the decline of violent crime in the United States since the early 1990s.

Exposure Routes

Lead is a common environmental pollutant. Causes of environmental contamination include industrial use of lead, such as is found in facilities that process lead-acid batteries or produce lead wire or pipes, and metal recycling and foundries. Children living near facilities that process lead, such as lead smelters, have been found to have unusually high blood lead levels. In August 2009, parents rioted in China after lead poisoning was found in nearly 2000 children living near zinc and manganese smelters. Lead exposure can occur from contact with lead in air, household dust, soil, water, and commercial products. Leaded gasoline has also been linked to increases in lead pollution. Some research has suggested a link between leaded gasoline and crime rates.

Occupational Exposure

In adults, occupational exposure is the main cause of lead poisoning. People can be exposed when working in facilities that produce a variety of lead-containing products; these include radiation shields, ammunition, certain surgical equipment, developing dental x-ray films prior to digital x-rays (each film packet had a lead liner to prevent the radiation from going through), fetal monitors, plumbing, circuit boards, jet engines, and ceramic glazes. In addition, lead miners and smelters, plumbers and fitters, auto mechanics, glass manufacturers, construction workers, battery manufacturers and recyclers, firing range instructors, and plastic manufacturers are at risk for lead exposure. Other occupations that present lead exposure risks include welding, manufacture of rubber, printing, zinc and copper smelting, processing of ore, combustion of solid waste, and production of paints and pigments. Parents who are exposed to lead in the workplace can bring lead dust home on clothes or skin and expose their children.

Battery recycling workers are at risk for lead exposure. This worker ladles molten lead into billets in a lead-acid battery recovery facility.

Paint

Some lead compounds are colorful and are used widely in paints, and lead paint is a major route of lead exposure in children. A study conducted in 1998–2000 found that 38 million housing units in the US had lead-based paint, down from a 1990 estimate of 64 million. Deteriorating lead paint can produce dangerous lead levels in household dust and soil. Deteriorating lead paint and lead-containing household dust are the main causes of chronic lead poisoning. The lead breaks down into the dust and since children are more prone to crawling on the floor, it is easily ingested. Many young children display pica, eating things that are not food. Even a small amount of a lead-containing product such as a paint chip or a sip of glaze can contain tens or hundreds of milligrams of lead. Eating chips of lead paint presents a particular hazard to children, generally producing more severe poisoning than occurs from dust. Because removing lead paint from dwellings, e.g. by sanding or torching creates lead-containing dust and fumes, it is generally safer to seal the lead paint under new paint (excepting moveable windows and doors, which create paint dust when operated). Alternately, special precautions must be taken if the lead paint is to be removed. In oil painting it was once common for colours such as yellow or white to be made with lead carbonate. Lead white oil colour was the main white of oil painters until superseded by compounds containing zinc or titanium in the mid-20th century. It is speculated that the painter Caravaggio and possibly Francisco Goya and Vincent Van Gogh had lead poisoning due to overexposure or carelessness when handling this colour.

Soil

Residual lead in soil contributes to lead exposure in urban areas. It has been thought that the more polluted an area is with various contaminants, the more likely it is to contain lead. However, this is not always the case, as there are several other reasons for lead contamination in soil. Lead content in soil may be caused by broken-down lead paint, residues from lead-containing gasoline, used engine oil, or pesticides used in the past, contaminated landfills, or from nearby industries such as foundries or smelters. Although leaded soil is less of a problem in countries that no longer have leaded gasoline, it remains prevalent, raising concerns about the safety of urban agriculture; eating food grown in contaminated soil can present a lead hazard.

A lead warning on a fuel pump. Tetraethyllead, which used to be added to automotive gasoline (and still is added to some aviation gasolines), contributed to soil contamination.

Water

Lead from the atmosphere or soil can end up in groundwater and surface water. It is also potentially in drinking water, e.g. from plumbing and fixtures that are either made of lead or have lead solder. Since acidic water breaks down lead in plumbing more readily, chemicals can be added to municipal water to increase the pH and thus reduce the corrosivity of the public water supply. Chloramines, which were adopted as a substitute for chlorine disinfectants due to fewer health concerns, increase corrositivity. In the US, 14–20% of total lead exposure is attributed to drinking water. In 2004, a team of seven reporters from *The Washington Post* discovered high levels of lead in the drinking water in Washington, D.C. and won an award for investigative reporting for a series of articles about this contamination. In the Flint water crisis, a switch to a more corrosive municipal water source elevated lead levels in drinking water in domestic tap water.

In Australia, collecting rainwater from roof runoff used as potable water may contain lead if there are lead contaminants on the roof or in the storage tank. The Australian Drinking Water Guidelines allow a maximum of .01 mg/L lead in water.

Lead-Containing Products

Lead can be found in products such as kohl, an ancient cosmetic from the Middle East, South Asia, and parts of Africa that has many names; and from some toys. In 2007, millions of toys made in China were recalled from multiple countries owing to safety hazards including lead paint. Vinyl mini-blinds, found especially in older housing, may contain lead. Lead is commonly incorporated into herbal remedies such as Indian Ayurvedic preparations and remedies of Chinese origin. There are also risks of elevated blood lead levels caused by folk remedies like *azarcon* and *greta*, which each contain about 95% lead.

Ingestion of metallic lead, such as small lead fishing lures, increases blood lead levels and can be fatal. Ingestion of lead-contaminated food is also a threat. Ceramic glaze often contains lead, and dishes that have been improperly fired can leach the metal into food, potentially causing severe poisoning. In some places, the solder in cans used for food contains lead. When manufacturing medical instruments and hardware, solder containing lead may be present. People who eat animals hunted with lead bullets may be at risk for lead exposure. Bullets lodged in the body rarely cause significant levels of lead, but bullets lodged in the joints are the exception, as they deteriorate and release lead into the body over time.

In May 2015, Indian food safety regulators in the state of Uttar Pradesh found that samples of Maggi 2 Minute Noodles contained lead up to 17 times beyond permissible limits. On 3 June 2015, New Delhi Government banned the sale of Maggi noodles in New Delhi stores for 15 days because it was found to contain lead beyond the permissible limit. The Gujarat FDA on June 4, 2015 banned the noodles for 30 days after 27 out of 39 samples were detected with objectionable levels of metallic lead, among other things. Some India's biggest retailers like Future Group, Big Bazaar, Easyday and Nilgiris have imposed a nationwide ban on Maggi noodles. Many other states too have banned Maggi noodles.

Bullets

Contact with ammunition is a source of lead exposure. As of 2013, lead-based ammunition production is the second largest annual use of lead in the US, accounting for over 60,000 metric tons consumed in 2012, second only to the manufacture of storage batteries. The Environmental Protection Agency (EPA) does not regulate lead, as a matter of law. Lead birdshot is banned in some areas, but this is primarily for the benefit of the birds and their predators, rather than humans. Non-lead alternatives include steel, tungsten-nickel-iron, bismuth-tin, and tungsten-polymer.

Because game animals can be shot using lead bullets, the potential for lead ingestion from game meat consumption has been studied clinically and epidemiologically. In a recent study conducted by the CDC, a cohort from North Dakota was enrolled and asked to self-report historical consumption of game meat, and participation in other activities that could cause lead exposure. The study found that participants' age, sex, housing age, current hobbies with potential for lead exposure, and game consumption were all associated with blood lead level (PbB).

This study has been cited by popular media as simple evidence that hunting increases exposure to lead poisoning, prompting the University of Illinois Extension to release a statement that there is no such risk. Concerning the CDC report, the authors' conclusion in a related Epi-AID Trip Report notes the small increase associated with game consumption in the study, and urges interpretation with respect to environmental context:

While this study suggests that consumption of wild game meat can adversely affect PbB, no participant had PbB higher than the CDC recommended threshold of 10µg/dl—the level at which CDC recommends case management; and the geometric mean PbB among this study population (1.17µg/dl) was lower than the overall population geometric mean PbB in the United States (1.60 µg/dl). The clinical significance of low PbB in this sample population and the small quantitative increase of 0.30µg/dl in PbB associated with wild game consumption should be interpreted in the context of naturally occurring PbB.

Copper-jacketed, lead-based bullets are more economical to produce and use than lead or any other material. Alternative materials are available such as steel, copper, and tungsten, but alternatives are universally less effective and/or more expensive. However, the biggest impediment to using the vast majority of alternatives relates to current laws in the United States pertaining to armor-piercing rounds. Laws and regulations relating to armor-piercing ammunition expressly prohibit the use of brass, bronze, steel, tungsten, and nearly every metallic alternative in any bullet that can be shot by a handgun, which at this time is nearly every caliber smaller than 50BMG (including the popular .223 Remington, .308 Winchester and .30-06 to name just a few). Some lead-based bullets are resistant to fragmentation, offering hunters the ability to clean game animals with negligible risk of including lead fragments in prepared meat. Other bullets are prone to fragmentation and exacerbate the risk of lead ingestion from prepared meat. In practice, use of a non-fragmenting bullet and proper cleaning of the game animal's wound can eliminate the risk of lead ingestion from eating game; however, isolating such practice to experimentally determine its association with blood lead levels in study is difficult. Bismuth is an element used as a lead-replacement for shotgun pellets used in waterfowl hunting although shotshells made from bismuth are nearly ten times the cost of lead.

Pathophysiology

Exposure occurs through inhalation, ingestion or occasionally skin contact. Lead may be taken in through direct contact with mouth, nose, and eyes (mucous membranes), and through breaks in the skin. Tetraethyllead, which was a gasoline additive and is still used in fuels such as aviation fuel, passes through the skin; however inorganic lead found in paint, food, and most lead-containing consumer products is only minimally absorbed through the skin. The main sources of absorption of inorganic lead are from ingestion and inhalation. In adults, about 35–40% of inhaled lead dust is deposited in the lungs, and about 95% of that goes into the bloodstream. Of ingested inorganic lead, about 15% is absorbed, but this percentage is higher in children, pregnant women, and people with deficiencies of calcium, zinc, or iron. Children and infants may absorb about 50% of ingested lead, but little is known about absorption rates in children.

The main body compartments that store lead are the blood, soft tissues, and bone; the half-life of lead in these tissues is measured in weeks for blood, months for soft tissues, and years for bone. Lead in the bones, teeth, hair, and nails is bound tightly and not available to other tissues, and is generally thought not to be harmful. In adults, 94% of absorbed lead is deposited in the bones and teeth, but children only store 70% in this manner, a fact which may partially account for the more serious health effects on children. The estimated half-life of lead in bone is 20 to 30 years, and bone can introduce lead into the bloodstream long after the initial exposure is gone. The half-life of lead in the blood in men is about 40 days, but it may be longer in children and pregnant women, whose bones are undergoing remodeling, which allows the lead to be continuously re-introduced into the bloodstream. Also, if lead exposure takes place over years, clearance is much slower, partly due to the re-release of lead from bone. Many other tissues store lead, but those with the highest concentrations (other than blood, bone, and teeth) are the brain, spleen, kidneys, liver, and lungs. It is removed from the body very slowly, mainly through urine. Smaller amounts of lead are also eliminated through the feces, and very small amounts in hair, nails, and sweat.

Lead has no known physiologically relevant role in the body, and its harmful effects are myriad. Lead and other heavy metals create reactive radicals which damage cell structures including DNA and cell membranes. Lead also interferes with DNA transcription, enzymes that help in the synthesis of vitamin D, and enzymes that maintain the integrity of the cell membrane. Anemia may result when the cell membranes of red blood cells become more fragile as the result of damage to their membranes. Lead interferes with metabolism of bones and teeth and alters the permeability of blood vessels and collagen synthesis. Lead may also be harmful to the developing immune system, causing production of excessive inflammatory proteins; this mechanism may mean that lead exposure is a risk factor for asthma in children. Lead exposure has also been associated with a decrease in activity of immune cells such as polymorphonuclear leukocytes. Lead also interferes with the normal metabolism of calcium in cells and causes it to build up within them.

Enzymes

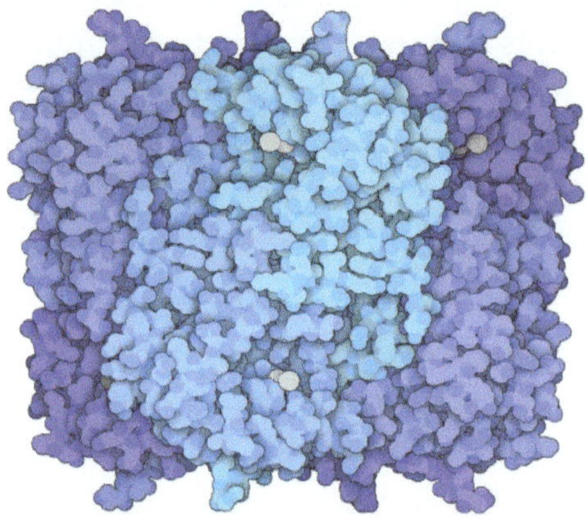

ALAD enzyme with lead bound (PDB: 1QNV)

The primary cause of lead's toxicity is its interference with a variety of enzymes because it binds to sulfhydryl groups found on many enzymes. Part of lead's toxicity results from its ability to mimic other metals that take part in biological processes, which act as cofactors in many enzymatic reactions, displacing them at the enzymes on which they act. Lead is able to bind to and interact with many of the same enzymes as these metals but, due to its differing chemistry, does not properly function as a cofactor, thus interfering with the enzyme's ability to catalyze its normal reaction or reactions. Among the essential metals with which lead interacts are calcium, iron, and zinc.

One of the main causes for the pathology of lead is that it interferes with the activity of an essential enzyme called delta-aminolevulinic acid dehydratase, or ALAD (see image of the enzyme structure), which is important in the biosynthesis of heme, the cofactor found in hemoglobin. Lead also inhibits the enzyme ferrochelatase, another enzyme involved in the formation of heme. Ferrochelatase catalyzes the joining of protoporphyrin and Fe^{2+} to form heme. Lead's interference with heme synthesis results in production of zinc protoporphyrin and the development of anemia. Another effect of lead's interference with heme synthesis is the buildup of heme precursors, such as aminolevulinic acid, which may be directly or indirectly harmful to neurons.

Neurons

The brain is the organ most sensitive to lead exposure. Lead is able to pass through the endothelial cells at the blood brain barrier because it can substitute for calcium ions and be uptaken by Calcium-ATPase pumps. Lead poisoning interferes with the normal development of a child's brain and nervous system; therefore children are at greater risk of lead neurotoxicity than adults are. In a child's developing brain, lead interferes with synapse formation in the cerebral cortex, neurochemical development (including that of neurotransmitters), and organization of ion channels. It causes loss of neurons' myelin sheaths, reduces numbers of neurons, interferes with neurotransmission, and decreases neuronal growth.

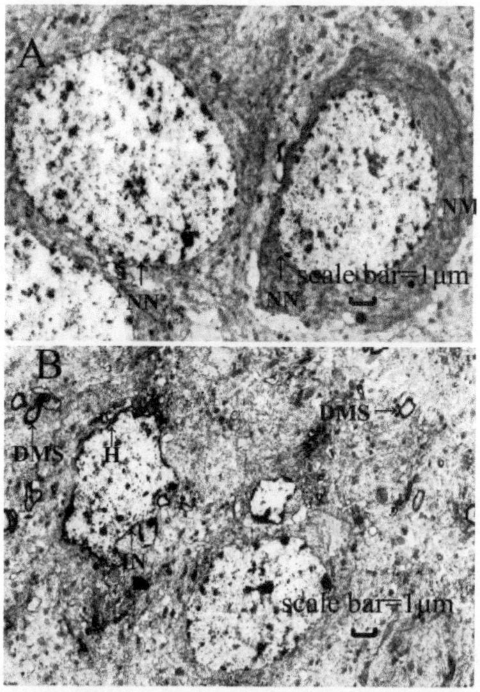

Lead exposure damages cells in the hippocampus, a part of the brain involved in memory. Hippocampi of lead-exposed rats (bottom) show structural damage such as irregular nuclei (IN) and denaturation of myelin (DNS) compared to controls (top).

Lead interferes with the release of neurotransmitters, chemicals used by neurons to send signals to other cells. It interferes with the release of glutamate, a neurotransmitter important in many functions including learning, by blocking NMDA receptors. The targeting of NMDA receptors is thought to be one of the main causes for lead's toxicity to neurons. A Johns Hopkins University report found that in addition to inhibiting the NMDA receptor, lead exposure decreased the level of expression of the gene for the receptor in part of the brain. In addition, lead has been found in animal studies to cause programmed cell death in brain cells.

Diagnosis

Diagnosis includes determining the clinical signs and the medical history, with inquiry into possible routes of exposure. Clinical toxicologists, medical specialists in the area of poisoning, may be involved in diagnosis and treatment. The main tool in diagnosing and assessing the severity of lead poisoning is laboratory analysis of the blood lead level (BLL).

Blood film examination may reveal basophilic stippling of red blood cells (dots in red blood cells visible through a microscope), as well as the changes normally associated with iron-deficiency anemia (microcytosis and hypochromasia). However, basophilic stippling is also seen in unrelated conditions, such as megaloblastic anemia caused by vitamin B12 (colbalamin) and folate deficiencies.

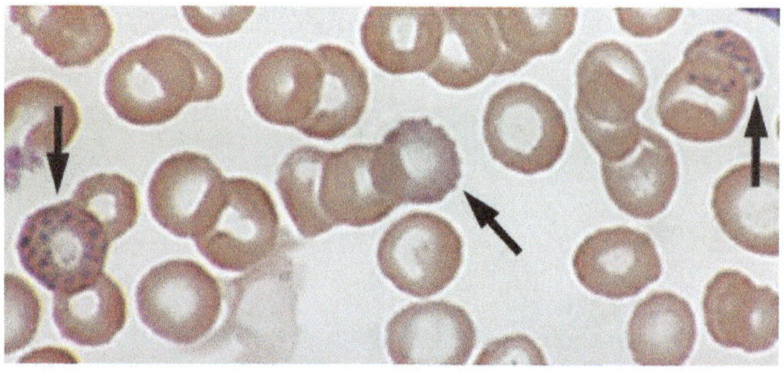

Basophilic stippling (arrows) of red blood cells in a 53-year-old who had elevated blood lead levels due to drinking repeatedly from glasses decorated with lead paint.

Exposure to lead also can be evaluated by measuring erythrocyte protoporphyrin (EP) in blood samples. EP is a part of red blood cells known to increase when the amount of lead in the blood is high, with a delay of a few weeks. Thus EP levels in conjunction with blood lead levels can suggest the time period of exposure; if blood lead levels are high but EP is still normal, this finding suggests exposure was recent. However, the EP level alone is not sensitive enough to identify elevated blood lead levels below about 35 μg/dL. Due to this higher threshold for detection and the fact that EP levels also increase in iron deficiency, use of this method for detecting lead exposure has decreased.

Blood lead levels are an indicator mainly of recent or current lead exposure, not of total body burden. Lead in bones can be measured noninvasively by X-ray fluorescence; this may be the best measure of cumulative exposure and total body burden. However this method is not widely available and is mainly used for research rather than routine diagnosis. Another radiographic sign of elevated lead levels is the presence of radiodense lines called lead lines at the metaphysis in the long bones of growing children, especially around the knees. These lead lines, caused by increased calcification due to disrupted metabolism in the growing bones, become wider as the duration of lead exposure increases. X-rays may also reveal lead-containing foreign materials such as paint chips in the gastrointestinal tract.

Fecal lead content that is measured over the course of a few days may also be an accurate way to estimate the overall amount of childhood lead intake. This form of measurement may serve as a useful way to see the extent of oral lead exposure from all the diet and environmental sources of lead.

Lead poisoning shares symptoms with other conditions and may be easily missed. Conditions that present similarly and must be ruled out in diagnosing lead poisoning include carpal tunnel syndrome, Guillain-Barré syndrome, renal colic, appendicitis, encephalitis in adults, and viral gastroenteritis in children. Other differential diagnoses in children include constipation, abdominal colic, iron deficiency, subdural hematoma, neoplasms of the central nervous system, emotional and behavior disorders, and intellectual disability.

Reference Levels

The current reference range for acceptable blood lead concentrations in healthy persons without excessive exposure to environmental sources of lead is less than 5 μg/dL for children. It was less than 25 μg/dL for adults. Previous to 2012 the value for children was 10 (μg/dl). The current biological exposure index (a level that should not be exceeded) for lead-exposed workers in the U.S. is 30 μg/dL in a random blood specimen.

The National Institute for Occupational Safety and Health (CDC/NIOSH) reference blood lead level in adults is 10 µg/dL The U.S. national BLL geometric mean among adults was 1.2 µg/dL in 2009–2010.

Blood lead concentrations in poisoning victims have ranged from 30->80 µg/dL in children exposed to lead paint in older houses, 77–104 µg/dL in persons working with pottery glazes, 90–137 µg/dL in individuals consuming contaminated herbal medicines, 109–139 µg/dL in indoor shooting range instructors and as high as 330 µg/dL in those drinking fruit juices from glazed earthenware containers.

Prevention

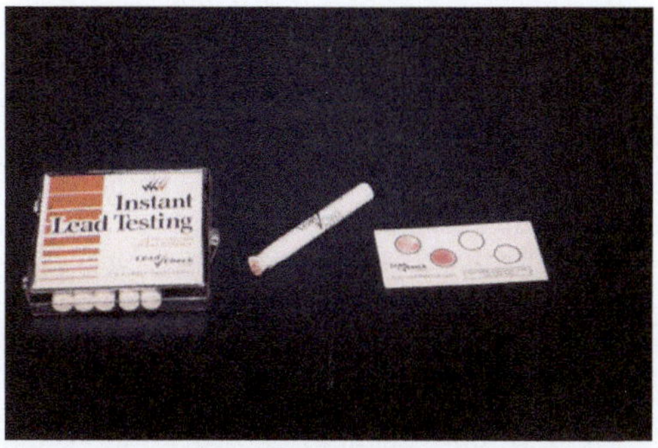

Testing kits are commercially available for detecting lead. These swabs, when wiped on a surface, turn red in the presence of lead.

In most cases, lead poisoning is preventable by avoiding exposure to lead. Prevention strategies can be divided into individual (measures taken by a family), preventive medicine (identifying and intervening with high-risk individuals), and public health (reducing risk on a population level).

Recommended steps by individuals to reduce the blood lead levels of children include increasing their frequency of hand washing and their intake of calcium and iron, discouraging them from putting their hands to their mouths, vacuuming frequently, and eliminating the presence of lead-containing objects such as blinds and jewellery in the house. In houses with lead pipes or plumbing solder, these can be replaced. Less permanent but cheaper methods include running water in the morning to flush out the most contaminated water, or adjusting the water's chemistry to prevent corrosion of pipes. Lead testing kits are commercially available for detecting the presence of lead in the household. As hot water is more likely than cold water to contain higher amounts of lead, use only cold water from the tap for drinking, cooking, and for making baby formula. Since most of the lead in household water usually comes from plumbing in the house and not from the local water supply, using cold water can avoid lead exposure. Measures such as dust control and household education do not appear to be effective in changing children's blood levels.

Screening is an important method in preventive medicine strategies. Screening programs exist to test the blood of children at high risk for lead exposure, such as those who live near lead-related industries.

Prevention measures also exist on national and municipal levels. Recommendations by health professionals for lowering childhood exposures include banning the use of lead where it is not essential and strengthening regulations that limit the amount of lead in soil, water, air, household dust, and products. Regulations exist to limit the amount of lead in paint; for example, a 1978 law in the US restricted the lead in paint for residences, furniture, and toys to 0.06% or less. In October 2008, the US Environmental Protection Agency reduced the allowable lead level by a factor of ten to 0.15 micrograms per cubic meter of air, giving states five years to comply with the standards. The European Union's Restriction of Hazardous Substances Directive limits amounts of lead and other toxic substances in electronics and electrical equipment. In some places, remediation programs exist to reduce the presence of lead when it is found to be high, for example in drinking water. As a more radical solution, entire towns located near former lead mines have been "closed" by the government, and the population resettled elsewhere, as was the case with Picher, Oklahoma in 2009.

Treatment

CDC management guidelines for children with elevated blood levels	
Blood lead level (μg/dL)	Treatment
10–14	Education, repeat screening
15–19	Repeat screening, case management to abate sources
20–44	Medical evaluation, case management
45–69	Medical evaluation, chelation, case management
>69	Hospitalization, immediate chelation, case management

The mainstays of treatment are removal from the source of lead and, for people who have significantly high blood lead levels or who have symptoms of poisoning, chelation therapy. Treatment of iron, calcium, and zinc deficiencies, which are associated with increased lead absorption, is another part of treatment for lead poisoning. When lead-containing materials are present in the gastrointestinal tract (as evidenced by abdominal X-rays), whole bowel irrigation, cathartics, endoscopy, or even surgical removal may be used to eliminate it from the gut and prevent further exposure. Lead-containing bullets and shrapnel may also present a threat of further exposure and may need to be surgically removed if they are in or near fluid-filled or synovial spaces. If lead encephalopathy is present, anticonvulsants may be given to control seizures, and treatments to control swelling of the brain include corticosteroids and mannitol. Treatment of organic lead poisoning involves removing the lead compound from the skin, preventing further exposure, treating seizures, and possibly chelation therapy for people with high blood lead concentrations.

A chelating agent is a molecule with at least two negatively charged groups that allow it to form complexes with metal ions with multiple positive charges, such as lead. The chelate that is thus formed is nontoxic and can be excreted in the urine, initially at up to 50 times the normal rate. The chelating agents used for treatment of lead poisoning are edetate disodium calcium ($CaNa_2EDTA$), dimercaprol (BAL), which are injected, and succimer and d-penicillamine, which are administered orally. Chelation therapy is used in cases of acute lead poisoning, severe poisoning, and encephalopathy, and is considered for people with blood lead levels above 25 µg/dL. While the use of chelation for people with symptoms of lead poisoning is widely supported, use in asymptomatic people with high blood lead levels is more controversial. Chelation therapy is of limited value for cases of chronic exposure to low levels of lead. Chelation therapy is usually stopped when symptoms resolve or when blood lead levels return to premorbid levels. When lead exposure has taken place over a long period, blood lead levels may rise after chelation is stopped because lead is leached into blood from stores in the bone; thus repeated treatments are often necessary.

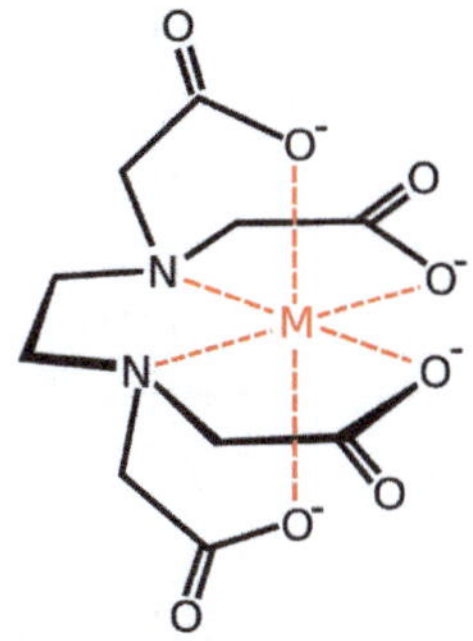

EDTA, a chelating agent, binds a heavy metal, sequestering it.

People receiving dimercaprol need to be assessed for peanut allergies since the commercial formulation contains peanut oil. Calcium EDTA is also effective if administered four hours after the administration of dimercaprol. Administering dimercaprol, DMSA (Succimer), or DMPS prior to calcium EDTA is necessary to prevent the redistribution of lead into the central nervous system. Dimercaprol used alone may also redistribute lead to the brain and testes. An adverse side effect of calcium EDTA is renal toxicity. Succimer (DMSA) is the preferred agent in mild to moderate lead poisoning cases. This may be the case in instances where children have a blood lead level >25µg/dL. The most reported adverse side effect for succimer is gastrointestinal disturbances. It is also important to note that chelation therapy only lowers blood lead levels and may not prevent the lead-induced cognitive problems associated with lower lead levels in tissue. This may be because of the inability of these agents to remove sufficient amounts of lead from tissue or inability to reverse preexisting damage. Chelating agents can have adverse effects; for example, chelation therapy can lower the body's levels of necessary nutrients like zinc. Chelating agents taken orally can increase the body's absorption of lead through the intestine.

Chelation challenge, also known as provocation testing, is used to indicate an elevated and mobilizable body burden of heavy metals including lead. This testing involves collecting urine before and after administering a one-off dose of chelating agent to mobilize heavy metals into the urine. Then urine is analyzed by a laboratory for levels of heavy metals; from this analysis overall body burden is inferred. Chelation challenge mainly measures the burden of lead in soft tissues, though whether it accurately reflects long-term exposure or the amount of lead stored in bone remains controversial. Although the technique has been used to determine whether chelation therapy is indicated and to diagnose heavy metal exposure, some evidence does not support these uses as blood levels after chelation are not comparable to the reference range typically used to diagnose heavy metal poisoning. The single chelation dose could also redistribute the heavy metals to more sensitive areas such as central nervous system tissue.

Epidemiology

Since lead has been used widely for centuries, the effects of exposure are worldwide. Environmental lead is ubiquitous, and everyone has some measurable blood lead level. Atmospheric lead pollution increased dramatically beginning in the 1950s as a result of the widespread use of leaded gasoline. Lead is one of the largest environmental medicine problems in terms of numbers of people exposed and the public health toll it takes. Lead exposure accounts for about 0.2% of all deaths and 0.6% of disability adjusted life years globally.

Although regulation reducing lead in products has greatly reduced exposure in the developed world since the 1970s, lead is still allowed in products in many developing countries. In all countries that have banned leaded gasoline, average blood lead levels have fallen sharply. However, some developing countries still allow leaded gasoline, which is the primary source of lead exposure in most developing countries. Beyond exposure from gasoline, the frequent use of pesticides in developing countries adds a risk of lead exposure and subsequent poisoning. Poor children in developing countries are at especially high risk for lead poisoning. Of North American children, 7% have blood lead levels above 10 μg/dL, whereas among Central and South American children, the percentage is 33 to 34%. About one fifth of the world's disease burden from lead poisoning occurs in the Western Pacific, and another fifth is in Southeast Asia.

In developed countries, people with low levels of education living in poorer areas are most at risk for elevated lead. In the US, the groups most at risk for lead exposure are the impoverished, city-dwellers, and immigrants. African-American children and those living in old housing have also been found to be at elevated risk for high blood lead levels in the US. Low-income people often live in old housing with lead paint, which may begin to peel, exposing residents to high levels of lead-containing dust.

Risk factors for elevated lead exposure include alcohol consumption and smoking (possibly because of contamination of tobacco leaves with lead-containing pesticides). Adults with certain risk factors might be more susceptible to toxicity; these include calcium and iron deficiencies, old age, disease of organs targeted by lead (e.g. the brain, the kidneys), and possibly genetic susceptibility. Differences in vulnerability to lead-induced neurological damage between males and females have also been found, but some studies have found males to be at greater risk, while others have found females to be.

In adults, blood lead levels steadily increase with increasing age. In adults of all ages, men have higher blood lead levels than women do. Children are more sensitive to elevated blood lead levels than adults are. Children may also have a higher intake of lead than adults; they breathe faster and may be more likely to have contact with and ingest soil. Children of ages one to three tend to have the highest blood lead levels, possibly because at that age they begin to walk and explore their environment, and they use their mouths in their exploration. Blood levels usually peak at about 18–24 months old. In many countries including the US, household paint and dust are the major route of exposure in children.

Notable Cases

Cases of mass lead poisoning can occur, primarily in underdeveloped countries.

15,000 people are being relocated from Jiyuan in central Henan province to other locations after 1000 children living around China's largest smelter plant (owned and operated by Yuguang Gold and Lead) were found to have excess lead in their blood. The total cost of this project is estimated to around 1 billion yuan ($150 million). 70% of the cost will be paid by local government and the smelter company, while the rest will be paid by the residents themselves. The government has suspended production at 32 of 35 lead plants. The affected area includes people from 10 different villages.

The Zamfara State lead poisoning epidemic occurred in Nigeria in 2010. As of October 5, 2010 at least 400 children have died from the effects of lead poisoning.

Prognosis

Reversibility

Outcome is related to the extent and duration of lead exposure. Effects of lead on the physiology of the kidneys and blood are generally reversible; its effects on the central nervous system are not. While peripheral effects in adults often go away when lead exposure ceases, evidence suggests that most of lead's effects on a child's central nervous system are irreversible. Children with lead poisoning may thus have adverse health, cognitive, and behavioral effects that follow them into adulthood.

Encephalopathy

Lead encephalopathy is a medical emergency and causes permanent brain damage in 70–80% of children affected by it, even those that receive the best treatment. The mortality rate for people who develop cerebral involvement is about 25%, and of those who survive who had lead encephalopathy symptoms by the time chelation therapy was begun, about 40% have permanent neurological problems such as cerebral palsy.

Long-Term

Exposure to lead may also decrease lifespan and have health effects in the long term. Death rates from a variety of causes have been found to be higher in people with elevated blood lead levels; these include cancer, stroke, and heart disease, and general death rates from all causes. Lead is considered a possible human carcinogen based on evidence from animal studies. Evidence also suggests that age-related mental decline and psychiatric symptoms are correlated with lead exposure. Cumulative exposure over a prolonged period may have a more important effect on some aspects of health than recent exposure. Some health effects, such as high blood pressure, are only significant risks when lead exposure is prolonged (over about one year).

History

Dioscorides noted lead's effect on the mind in the first century A.D.

Roman lead water pipes with taps

Lead poisoning was among the first known and most widely studied work and environmental hazards. One of the first metals to be smelted and used, lead is thought to have been discovered and first mined in Anatolia around 6500 BCE. Its density, workability, and corrosion-resistance were among the metal's attractions.

In the 2nd century BCE the Greek botanist Nicander described the colic and paralysis seen in lead-poisoned people. Dioscorides, a Greek physician who lived in the 1st century CE, wrote that lead makes the mind "give way".

Lead was used extensively in Roman aqueducts from about 500 BCE to 300 CE Julius Caesar's engineer, Vitruvius, reported, "water is much more wholesome from earthenware pipes than from lead pipes. For it seems to be made injurious by lead, because white lead is produced by it, and this is said to be harmful to the human body." Gout, prevalent in affluent Rome, is thought to be the result of lead, or leaded eating and drinking vessels. Sugar of lead (lead(II) acetate) was used to sweeten wine, and the gout that resulted from this was known as "saturnine" gout. It is even hypothesized that lead poisoning may have contributed to the decline of the Roman Empire, a hypothesis thoroughly disputed:

The great disadvantage of lead has always been that it is poisonous. This was fully recognised by the ancients, and Vitruvius specifically warns against its use. Because it was nevertheless used in profusion for carrying drinking water, the conclusion has often been drawn that the Romans must therefore have suffered from lead poisoning; sometimes conclusions are carried even further and it is inferred that this caused infertility and other unwelcome conditions, and that lead plumbing was largely responsible for the decline and fall of Rome.

Two things make this otherwise attractive hypothesis impossible. First, the calcium carbonate deposit that formed so thickly inside the aqueduct channels also formed inside the pipes, effectively insulating the water from the lead, so that the two never touched. Second, because the Romans had so few taps and the water was constantly running, it was never inside the pipes for more than a few minutes, and certainly not long enough to become contaminated.

However, recent research supports the idea that the lead found in the water came from the supply pipes, rather than another source of contamination. It was not unknown for locals to punch holes in the pipes to draw water off, increasing the number of people exposed to the lead.

Thirty years ago, Jerome Nriagu argued in a milestone paper that Roman civilization collapsed as a result of lead poisoning. Clair Patterson, the scientist who convinced governments to ban lead from gasoline, enthusiastically endorsed this idea, which nevertheless triggered a volley of publications aimed at refuting it. Although today lead is no longer seen as the prime culprit of Rome's demise, its status in the system of water distribution by lead pipes (fistulæ) still stands as a major public health issue. By measuring Pb isotope compositions of sediments from the Tiber River and the Trajanic Harbor, the present work shows that "tap water" from ancient Rome had 100 times more lead than local spring waters.

Romans also consumed lead through the consumption of defrutum, carenum, and sapa, musts made by boiling down fruit in lead cookware. Defrutum and its relatives were used in Ancient Roman cuisine and cosmetics, including as a food preservative. The use of leaden cookware, though popular, was not the general standard and copper cookware was used far more generally. There is also no indication how often sapa was added or in what quantity.

The consumption of sapa as having a role in the fall of the Roman Empire was used in a theory proposed by geochemist Jerome Nriagu to state that "lead poisoning contributed to the decline of the Roman Empire". In 1984, John Scarborough, a pharmacologist and classicist, criticized the conclusions drawn by Nriagu's book as "so full of false evidence, miscitations, typographical errors, and a blatant flippancy regarding primary sources that the reader cannot trust the basic arguments."

After antiquity, mention of lead poisoning was absent from medical literature until the end of the Middle Ages. In 1656 the German physician Samuel Stockhausen recognized dust and fumes containing lead compounds as the cause of disease, called since ancient Roman times *morbi metallici*, that were known to afflict miners, smelter workers, potters, and others whose work exposed them to the metal.

The painter Caravaggio might have died of lead poisoning. Bones with high lead levels were recently found in a grave thought likely to be his. Paints used at the time contained high amounts of lead salts. Caravaggio is known to have exhibited violent behavior, as caused by lead poisoning.

In 17th-century Germany, the physician Eberhard Gockel discovered lead-contaminated wine to be the cause of an epidemic of colic. He had noticed that monks who did not drink wine were healthy, while wine drinkers developed colic, and traced the cause to sugar of lead, made by simmering litharge with vinegar. As a result, Eberhard Ludwig, Duke of Württemberg issued an edict in 1696 banning the adulteration of wines with litharge.

In the 18th century lead poisoning was fairly frequent on account of the widespread drinking of rum, which was made in stills with a lead component (the "worm"). It was a significant cause of mortality amongst slaves and sailors in the colonial West Indies. Lead poisoning from rum was also noted in Boston. Benjamin Franklin suspected lead to be a risk in 1786. Also in the 18th century, "Devonshire colic" was the name given to the symptoms suffered by people of Devon who drank cider made in presses that were lined with lead. Lead was added to cheap wine illegally in the 18th and early 19th centuries as a sweetener. The composer Beethoven, a heavy wine drinker, suffered elevated lead levels (as later detected in his hair) possibly due to this; the cause of his death is controversial, but lead poisoning is a contender as a factor.

With the Industrial Revolution in the 19th century, lead poisoning became common in the work setting. The introduction of lead paint for residential use in the 19th century increased childhood exposure to lead; for millennia before this, most lead exposure had been occupational. An important step in the understanding of childhood lead poisoning occurred when toxicity in children from lead paint was recognized in Australia in 1897. France, Belgium, and Austria banned white lead interior paints in 1909; the League of Nations followed suit in 1922. However, in the United States, laws banning lead house paint were not passed until 1971, and it was phased out and not fully banned until 1978.

The 20th century saw an increase in worldwide lead exposure levels due to the increased widespread use of the metal. Beginning in the 1920s, lead was added to gasoline to improve its combustion; lead from this exhaust persists today in soil and dust in buildings. Blood lead levels worldwide have been declining sharply since the 1980s, when leaded gasoline began to be phased out. In those countries that have banned lead in solder for food and drink cans and have banned leaded gasoline additives, blood lead levels have fallen sharply since the mid-1980s.

The levels found today in most people are orders of magnitude greater than those of pre-industrial society. Due to reductions of lead in products and the workplace, acute lead poisoning is rare in most countries today, but low level lead exposure is still common. It was not until the second half of the 20th century that subclinical lead exposure became understood to be a problem. During the end of the 20th century, the blood lead levels deemed acceptable steadily declined. Blood lead levels once considered safe are now considered hazardous, with no known safe threshold.

In the 1980s Herbert Needleman was falsely accused of scientific misconduct by the Lead Industry Associates.

In 2002 Tommy Thompson, secretary of Health and Human Services appointed at least two persons with conflicts of interest to the CDC's Lead Advisory Committee.

In 2014 a case by the state of California against a number of companies decided against Sherwin-Williams NL Industries and ConAgra and required them to pay $1.15 billion.

Studies have found a weak link between lead from leaded gasoline and crime rates.

Other Species

Humans are not alone in suffering from lead's effects; plants and animals are also affected by lead toxicity to varying degrees depending on species. Animals experience many of the same effects of lead exposure as humans do, such as abdominal pain, peripheral neuropathy, and behavioral changes such as increased aggression. Much of what is known about human lead toxicity and its effects is derived from animal studies. Animals are used to test the effects of treatments, such as chelating agents, and to provide information on the pathophysiology of lead, such as how it is absorbed and distributed in the body.

Farm animals such as cows and horses as well as pet animals are also susceptible to the effects of lead toxicity. Sources of lead exposure in pets can be the same as those that present health threats to humans sharing the environment, such as paint and blinds, and there is sometimes lead in toys made for pets. Lead poisoning in a pet dog may indicate that children in the same household are at increased risk for elevated lead levels.

Wildlife

Turkey vultures, *Cathartes aura* (shown), and California condors can be poisoned when they eat carcasses of animals shot with lead pellets.

Lead, one of the leading causes of toxicity in waterfowl, has been known to cause die-offs of wild bird populations. When hunters use lead shot, waterfowl such as ducks can ingest the spent pellets later and be poisoned; predators that eat these birds are also at risk. Lead shot-related waterfowl poisonings were first documented in the US in the 1880s. By 1919, the spent lead pellets from waterfowl hunting was positively identified as the source of waterfowl deaths. Lead shot has been banned for hunting waterfowl in several countries, including the US in 1991 and Canada in 1997. Other threats to wildlife include lead paint, sediment from lead mines and smelters, and lead weights from fishing lines. Lead in some fishing gear has been banned in several countries.

The critically endangered California condor has also been affected by lead poisoning. As scavengers, condors eat carcasses of game that have been shot but not retrieved, and with them the fragments from lead bullets; this increases their lead levels. Among condors around the Grand Canyon, lead poisoning due to eating lead shot is the most frequently diagnosed cause of death. In an effort to protect this species, in areas designated as the California condor's range the use of projectiles containing lead has been banned to hunt deer, feral pigs, elk, pronghorn antelope, coyotes, ground squirrels, and other non-game wildlife. Also, conservation programs exist which routinely capture condors, check their blood lead levels, and treat cases of poisoning.

Melamine

Melamine is an organic base and a trimer of cyanamide, with a 1,3,5-triazine skeleton. Like cyanamide, it contains 67% nitrogen by mass and, if mixed with resins, has fire retardant properties due to its release of nitrogen gas when burned or charred, and has several other industrial uses. Melamine is also a metabolite of cyromazine, a pesticide. It is formed in the body of mammals that have ingested cyromazine. It has been reported that cyromazine can also be converted to melamine in plants.

Melamine combines with cyanuric acid and related compounds to form melamine cyanurate and related crystal structures, which have been implicated as contaminants or biomarkers in Chinese protein adulterations.

Etymology

The German word *melamin* was coined by combining the words: melam (a derivative of ammonium thiocyanate) and amine. Melamine is, therefore, unrelated etymologically to the root *melas*, from which the words melanin, a pigment, and melatonin, a hormone, are formed.

Uses

Melamine is combined with formaldehyde to produce melamine resin, a very durable thermosetting plastic used in high pressure decorative laminates such as Formica, melamine dinnerware, laminate flooring, and dry erase boards.

Melamine foam is used as insulation, soundproofing material and in polymeric cleaning products, such as Magic Eraser.

Melamine is one of the major components in Pigment Yellow 150, a colorant in inks and plastics.

Melamine also enters the fabrication of melamine poly-sulfonate used as superplasticizer for making high-resistance concrete. Sulfonated melamine formaldehyde (SMF) is a polymer used as cement admixture to reduce the water content in concrete while increasing the fluidity and the workability of the mix during its handling and pouring. It results in concrete with a lower porosity and a higher mechanical strength, exhibiting an improved resistance to aggressive environments and a longer lifetime.

The use of melamine as fertilizer for crops had been envisaged during the 1950s and 1960s because of its high nitrogen content (2/3). However, melamine is much more expensive to produce than are other common nitrogen fertilizers, such as urea. To be effective as a fertilizer, it is essential that the plant nutrients are released or made available in a manner that matches the needs of the growing crop. The nitrogen mineralization process for melamine is extremely slow, making this product both economically and scientifically impractical for use as a fertilizer.

Melamine dinnerware

Melamine and its salts are used as fire-retardant additives in paints, plastics, and paper.

Melamine derivatives of arsenical drugs are potentially important in the treatment of African trypanosomiasis.

Melamine use as non-protein nitrogen (NPN) for cattle was described in a 1958 patent. In 1978, however, a study concluded that melamine "may not be an acceptable non-protein N source for ruminants" because its hydrolysis in cattle is slower and less complete than other nitrogen sources such as cottonseed meal and urea.

Melamine is sometimes illegally added to food products in order to increase the apparent protein content. Standard tests, such as the Kjeldahl and Dumas tests, estimate protein levels by measuring the nitrogen content, so they can be misled by adding nitrogen-rich compounds such as melamine. An instrument (SPRINT) developed by the company CEM Corp allows the determination of protein content directly in some applications; the instrument cannot be fooled by adding melamine in the sample.

Melamine is also used as a nitrogen and carbon source for N-doped carbon nanotube. N-CNTs can be prepared via Chemical Vapor Deposition (CVD) method by pyrolysizing melamine under an Ar atmosphere in a horizontal glass tube. A thin film of iron (5 nm) is first deposited on a Si/SiO_2 wafer. N-CNT synthesis occurs at a furnace temperatures between 800 and 980 °C.

Toxicity

Melamine is described as "Harmful if swallowed, inhaled or absorbed through the skin. Chronic exposure may cause cancer or reproductive damage. Eye, skin and respiratory irritant". However, the short-term lethal dose is on a par with common table salt with an LD_{50} of more than 3 grams per kilogram of bodyweight. U.S. Food and Drug Administration (FDA) scientists explained that when melamine and cyanuric acid are absorbed into the bloodstream, they concentrate and interact in the urine-filled renal tubules, then crystallize and form large numbers of round, yellow crystals, which in turn block and damage the renal cells that line the tubes, causing the kidneys to malfunction.

The European Union set a standard for acceptable human consumption (tolerable daily intake, TDI) of melamine at 0.2 mg per kg of body mass, (previously 0.5 milligrams), Canada declared a limit of 0.35 mg and the US FDA's limit was put at 0.063 mg daily (previously 0.63 mg). The World Health Organization's food safety director estimated that the amount of melamine a person could stand per day without incurring a bigger health risk, the TDI, was 0.2 mg per kg of body mass.

Toxicity of melamine can be mediated by intestinal microbes. In culture, Klebsiella terrigena, which rarely colonizes mammalian intestines, was shown to convert melamine to cyanuric acid directly. Rats colonized by K. terrigena showed greater melamine-induced kidney damage compared to those not colonized.

Acute Toxicity

Melamine is reported to have an oral LD_{50} of 3248 mg/kg based on rat data. It is also an irritant when inhaled or in contact with the skin or eyes. The reported dermal LD_{50} is >1000 mg/kg for rabbits. A study by USSR researchers in the 1980s suggested that melamine cyanurate, commonly used as a fire retardant, could be more toxic than either melamine or cyanuric acid alone. For rats and mice, the reported LD_{50} for melamine cyanurate was 4.1 g/kg (given inside the stomach) and 3.5 g/kg (via inhalation), compared to 6.0 and 4.3 g/kg for melamine and 7.7 and 3.4 g/kg for cyanuric acid, respectively.

A toxicology study in animals conducted after recalls of contaminated pet food concluded that the combination of melamine and cyanuric acid in diet does lead to acute renal failure in cats. A 2008 study produced similar experimental results in rats and characterized the melamine and cyanuric acid in contaminated pet food from the 2007 outbreak. A 2010 study from Lanzhou University attributed renal failure in humans to uric acid stone accumulation after ingestion of melamine resulting in a rapid aggregation of metabolites such as cyanuric acid diamide (ammeline) and cyanuric acid. A 2013 study demonstrated that melamine can be metabolized to cyanuric acid by gut bacteria. In particular, *Klebsiella terrigena* was determined to be a factor in melamine toxicity. In culture, *K. terrigena* was shown to convert melamine to cyanuric acid directly. Rats colonized by *K. terrigena* showed greater melamine-induced kidney damage compared to those not colonized.

Chronic Toxicity

Ingestion of melamine may lead to reproductive damage, or bladder or kidney stones, which can lead to bladder cancer.

A study in 1953 reported that dogs fed 3% melamine for a year had the following changes in their urine: (1) reduced specific gravity, (2) increased output, (3) melamine crystalluria, and (4) protein and occult blood.

A survey commissioned by the American Association of Veterinary Laboratory Diagnosticians suggested that crystals formed in the kidneys when melamine combined with cyanuric acid, "don't dissolve easily. They go away slowly, if at all, so there is the potential for chronic toxicity."

Treatment of Urolithiasis

Fast diagnosis and treatment of acute obstructive urolithiasis may prevent the development of acute renal failure. Urine alkalinization and stone liberalization have been reported to be the most effective treatments in humans.

Regulation in Food and Feed

The United Nations' food standards body, Codex Alimentarius Commission, has set the maximum amount of melamine allowed in powdered infant formula to 1 mg/kg and the amount of the chemical allowed in other foods and animal feed to 2.5 mg/kg. While not legally binding, the levels allow countries to ban importation of products with excessive levels of melamine.

Synthesis

Melamine was first synthesized by the German chemist Justus von Liebig in 1834. In early production, first calcium cyanamide is converted into dicyandiamide, then heated above its melting temperature to produce melamine. However, today most industrial manufacturers use urea in the following reaction to produce melamine:

$$6\,(NH_2)_2CO \rightarrow C_3H_6N_6 + 6\,NH_3 + 3\,CO_2$$

It can be understood as two steps.

First, urea decomposes into cyanic acid and ammonia in an endothermic reaction:

$$(NH_2)_2CO \rightarrow HCNO + NH_3$$

Then, cyanic acid polymerizes to form cyanuric acid which condenses with the liberated ammonia forming melamine which releases water which then reacts with cyanic acid present(which helps to drive the reaction) generating carbon dioxide and ammonia.

$$6\ HCNO + 3\ NH_3 \rightarrow C_3H_6N_6 + 3\ CO_2 + 3NH_3$$

The second reaction is exothermic but the overall process is endothermic.

The above reaction can be carried out by either of two methods: catalyzed gas-phase production or high pressure liquid-phase production. In one method, molten urea is introduced onto a fluidized bed with catalyst for reaction. Hot ammonia gas is also present to fluidize the bed and inhibit deammonization. The effluent then is cooled. Ammonia and carbon dioxide in the off-gas are separated from the melamine-containing slurry. The slurry is further concentrated and crystallized to yield melamine. Major manufacturers and licensors such as Orascom Construction Industries, BASF, and Eurotecnica have developed some proprietary methods.

The off-gas contains large amounts of ammonia. Therefore, melamine production is often integrated into urea production, which uses ammonia as feedstock.

Crystallization and washing of melamine generates a considerable amount of waste water, which is a pollutant if discharged directly into the environment. The waste water may be concentrated into a solid (1.5–5% of the weight) for easier disposal. The solid may contain approximately 70% melamine, 23% oxytriazines (ammeline, ammelide, and cyanuric acid), 0.7% polycondensates (melem, melam, and melon). In the Eurotecnica process, however, there is no solid waste and the contaminants are decomposed to ammonia and carbon dioxide and sent as off gas to the upstream urea plant; accordingly, the waste water can be recycled to the melamine plant itself or used as clean cooling water make-up.

Production in Mainland China

Between the late 1990s and early 2000s, both consumption and production of melamine grew considerably in mainland China. By early 2006, melamine production in mainland China is reported to be in “serious surplus”. Between 2002 and 2007, while the global melamine price remained stable, a steep increase in the price of urea (feedstock for melamine) has reduced the profitability of melamine manufacturing. Currently, China is the world’s largest exporter of melamine, while its domestic consumption still grows by 10% per year. However, reduced profit has already caused other joint melamine ventures to be postponed there.

Surplus melamine has been an adulterant for feedstock and milk in mainland China for several years now because it can make diluted or poor quality material appear to be higher in protein content by elevating the total nitrogen content detected by some simple protein tests. Actions taken in 2008 by the Government of China have reduced the practice of adulteration, with the goal of eliminating it. Court trials began in December 2008 for six people linked to the scandal and ended in January 2009 with two of the convicts being sentenced to death and executed.

Melamine Poisoning by Tainted Food

Melamine has been involved in several food recalls after the discovery of severe kidney damage to children and pets poisoned by melamine-adulterated food.

2007 Animal Feed Recalls

In 2007, a pet food recall was initiated by Menu Foods and other pet food manufacturers who had found their products had been contaminated and caused serious illnesses or deaths in some of the animals that had eaten them. In March 2007, the US Food and Drug Administration reported finding white granular melamine in the pet food, in samples of white granular wheat gluten imported from a single source in China, Xuzhou Anying Biologic Technology as well as in crystalline form in the kidneys and in urine of affected animals. Further vegetable protein imported from China was later implicated.

In April 2007, *The New York Times* reported that the addition of "melamine scrap" into fish and livestock feed to give the false appearance of a higher level of protein was an "open secret" in many parts of mainland China, reporting that this melamine scrap was being produced by at least one plant processing coal into melamine. Four days later, the New York Times reported that, despite the widely reported ban on melamine use in vegetable proteins in mainland China, at least some chemical manufacturers continued to report selling it for use in animal feed and in products for human consumption. Li Xiuping, a manager at Henan Xinxiang Huaxing Chemical in Henan Province, stated, "Our chemical products are mostly used for additives, not for animal feed. Melamine is mainly used in the chemical industry, but it can also be used in making cakes." Shandong Mingshui Great Chemical Group, the company reported by the New York Times as producing melamine from coal, produces and sells both urea and melamine but does not list melamine resin as a product.

Another recall incident in 2007 involved melamine which had been purposely added as a binder to fish and livestock feed manufactured in the United States. This was traced to suppliers in Ohio and Colorado.

2008 Chinese Outbreak

In September 2008, several companies, including Nestlé, were implicated in a scandal involving milk and infant formula which had been adulterated with melamine, leading to kidney stones and other renal failure, especially among young children. By December 2008, nearly 300,000 people had become ill, with more than 50,000 infant hospitalizations and six infant deaths. In a study published in the *New England Journal of Medicine*, it was reported that melamine exposure increased the incidence of urinary tract stones by seven times in children. Melamine may have been added to fool government protein content tests after water was added to fraudulently dilute the milk. Because of melamine's high nitrogen content (66% by mass versus approx. 10–12% for typical protein), it can cause the protein content of food to appear higher than the true value. Officials estimate that about 20 percent of the dairy companies tested in China sell products tainted with melamine. On January 22, 2009, three of those involved in the scandal (including one conditional sentence) were sentenced to death in a Chinese court.

In October 2008, "Select Fresh Brown Eggs" imported to Hong Kong from the Hanwei Group in Dalian in northeastern China, were found to be contaminated with nearly twice the legal limit of melamine. York Chow, the health secretary of Hong Kong, said he thought animal feeds might be the source of the contamination and announced that the Hong Kong Centre for Food Safety would henceforward be testing all mainland Chinese pork, farmed fish, animal feed, chicken meat, eggs, and offal products for melamine.

As of July 2010, Chinese authorities were still reporting some seizures of melamine-contaminated dairy product in some provinces, though it was unclear whether these new contaminations constituted wholly new adulterations or were the result of illegal reuse of material from the 2008 adulterations.

On characterization and treatment of urinary stones in affected infants, the New England Journal of Medicine printed an editorial in March 2009, along with reports on cases from Beijing, Hong Kong and Taipei.

Urinary calculi specimens were collected from 15 cases treated in Beijing and were analyzed as unknown objects for their components at Beijing Institute of Microchemistry using infrared spectroscopy, nuclear magnetic resonance, and high performance liquid chromatography. The result of the analysis showed that the calculus was composed of melamine and uric acid, and the molecular ratio of uric acid to melamine was around 2:1.

In a 2009 study of 683 children diagnosed in Beijing in 2008 with nephrolithiasis and 6,498 children without nephrolithiasis aged < 3 years, investigators found that in children exposed to melamine levels < 0.2 mg/kg per day, the risk for nephrolithiasis was 1.7 times higher than in those without melamine exposure, suggesting that the risk of melamine-induced nephrolithiasis in young children starts at a lower intake level than the levels recommended by the World Health Organization.

In a study published in 2010, researchers from Beijing University studying ultrasound images of infants who fell ill in the 2008 contamination found that while most children in a rural Chinese area recovered, 12 per cent still showed kidney abnormalities six months later. "The potential for long-term complications after exposure to melamine remains a serious concern," the report said. "Our results suggest a need for further follow-up of affected children to evaluate the possible long-term impact on health, including renal function." Another 2010 follow-up study from Lanzhou University attributed the uric acid stone accumulation after ingestion of melamine to a rapid aggradation of metabolites such as cyanuric acid diamide (ammeline) and cyanuric acid and reported that urine alkalinization and stone liberalization were the most effective treatments.

Until the 2007 pet food recalls, melamine had not routinely been monitored in food, except in the context of plastic safety or insecticide residue. This could be due to the previously assumed low toxicity of melamine, and the relatively expensive methods of detection.

Following the deaths of children in China from powdered milk in 2008, the Joint Research Centre (JRC) of the European in Belgium set up a website about methods to detect melamine. In May 2009, the JRC published the results of a study that benchmarked the ability of labs around the world to accurately measure melamine in food. The study concluded that the majority of labs can effectively detect melamine in food.

In October 2008, the U.S. Food and Drug Administration (FDA) issued new methods for the analysis of melamine and cyanuric acid in infant formulations in the Laboratory Information Bulletin No 4421. Similar recommendations have been issued by other authorities, like the Japanese Ministry of Health, Labor and Welfare, both based on liquid chromatography – mass spectrometry (LC/MS) detection after hydrophilic interaction liquid chromatography (HILIC) separation.

The existing methods for melamine determination using a triple quadrupole liquid chromatography – mass spectrometry (LC/MS) after solid phase extraction (SPE) are often complex and time consuming. However, electrospray ionization methods coupled with mass spectrometry allow a rapid and direct analysis of samples with complex matrices: the native liquid samples are directly ionized under ambient conditions in their original solution. In December 2008, two new fast and inexpensive methods for detecting melamine in liquids have been published on-line in the *Chem. Comm. Journal* of the Royal Society of Chemistry (UK).

Ultrasound-assisted extractive electrospray ionization mass spectrometry (EESI-MS) has been developed at ETH Zurich (Switzerland) by Zhu *et al.*, (2008) for a rapid detection of melamine in untreated food samples. Ultrasounds are used to nebulize the melamine-containing liquids into a fine spray. The spray is then ionised by extractive electrospray ionisation (EESI) and analysed using tandem mass spectrometry (MS/MS). An analysis requires 30 seconds per sample. The limit of detection of melamine is a few nanograms of melamine per gram of milk.

Huang *et al.*, (2008) have also developed at Purdue University (US) a simpler instrumentation and a faster method by using a low-temperature plasma probe to ionize the samples. The major obstacles being solved, the ESI-MS technique allows now high-throughput analysis of melamine traces in complex mixtures.

The Melaminometer was a hypothetical design for a synthetic biology circuit, to be used for detecting melamine and related chemical analogues such as cyanuric acid. The conceptual project is hosted at OpenWetWare as open source biology in collaboration with DIYbio and has been discussed in various newspapers in the context of homebrew biotechnology. As of October 2009, the design has not been verified.

Because melamine resin is often used in food packaging and tableware, melamine at ppm level (1 part per million) in food and beverage has been reported due to migration from melamine-containing resins. Small amounts of melamine have also been reported in foodstuff as a metabolite product of cyromazine, an insecticide used on animals and crops.

The Food Safety and Inspection Service (FSIS) of the United States Department of Agriculture (USDA) provides a test method for analyzing cyromazine and melamine in animal tissues. In 2007, the FDA began using a high performance liquid chromatography test to determine the melamine, ammeline, ammelide, and cyanuric acid contamination in food. another procedure is based on surface-enhanced Raman spectroscopy (SERS).

Member States of the European Union are required under Commission Decision 2008/757/EC to ensure that all composite products containing at least 15% of milk product, originating from China, are systematically tested before import into the Community and that all such products which are shown to contain melamine in excess of 2.5 mg/kg are immediately destroyed.

Detection in Biological Specimens

Melamine has been determined in urine specimens from children who consumed adulterated milk products by liquid chromatography-mass spectrometry.

Melamine on Metal Surfaces

It is reported that melamine molecules adsorbed on Au(111) or Ag(111) surface tend to arrange into honeycomb or closed-packed structures. Such a self-assembly occurs due to the inter-molecular hydrogen bond interaction. This ordering was further investigated using classical Monte Carlo and DFT methods.

References

- Brunton, L.L.; Goodman, L.S.; Blumenthal, D.; Buxton, I.; Parker, K.L., eds. (2007). "Principles of toxicology". Goodman and Gilman's Manual of Pharmacology and Therapeutics. McGraw-Hill Professional. ISBN 0-07-144343-6.
- Casarett, LJ; Klaassen, CD; Doull, J, eds. (2007). "Toxic effects of metals". Casarett and Doull's Toxicology: The Basic Science of Poisons (7th ed.). McGraw-Hill Professional. ISBN 0-07-147051-4.
- Chisolm, J.J. (2004). "Lead poisoning". In Crocetti, M.; Barone, M.A.; Oski, F.A. Oski's Essential Pediatrics (2nd ed.). Lippincott Williams & Wilkins. ISBN 0-7817-3770-2.
- Dart, R.C.; Hurlbut, K.M.; Boyer-Hassen, L.V. (2004). "Lead". In Dart, RC. Medical Toxicology (3rd ed.). Lippincott Williams & Wilkins. ISBN 0-7817-2845-2.
- Grant, L.D. (2009). "Lead and compounds". In Lippmann, M. Environmental Toxicants: Human Exposures and Their Health Effects (3rd ed.). Wiley-Interscience. ISBN 0-471-79335-3.
- Henretig F.M. (2006). "Lead". In Goldfrank, LR. Goldfrank's Toxicologic Emergencies (8th ed.). McGraw-Hill Professional. ISBN 0-07-143763-0.
- Kosnett M.J. (2005). "Lead". In Brent, J. Critical Care Toxicology: Diagnosis and Management of the Critically Poisoned Patient. Gulf Professional Publishing. ISBN 0-8151-4387-7.
- Menkes, J.H. (2006). "Toxic and nutritional disorders". In Menkes, J.H.; Sarnat, H.B.; Maria, B.L. Child Neurology (7th ed.). Lippincott Williams & Wilkins. p. 706. ISBN 0-7817-5104-7.
- Olson, K.R. (2007). "Poisoning". In McPhee, S.J.; Tierney, L.M.; Papadakis, M.A. Current Medical Diagnosis and Treatment (46th ed.). McGraw-Hill Professional. ISBN 0-07-147247-9.
- Rambousek, A.J., ed. (2008). "The symptoms and treatment of industrial poisoning". Industrial Poisoning from Fumes, Gases, and Poisons of Manufacturing Processes. READ BOOKS. ISBN 1-4086-7025-9.
- Rubin, R.; Strayer, D.S., eds. (2008). "Environmental and nutritional pathology". Rubin's Pathology: Clinicopathologic Foundations of Medicine (5th ed.). Lippincott Williams & Wilkins. ISBN 0-7817-9516-8.
- Trevor, A.J.; Katzung, B.G.; Masters, S.B., eds. (2007). "Heavy metals". Katzung & Trevor's Pharmacology: Examination & Board Review (8th ed.). McGraw-Hill Professional. ISBN 0-07-148869-3.
- Yu, M.H. (2005). "Soil and water pollution: Environmental metals and metalloids". Environmental Toxicology: Biological and Health Effects of Pollutants. CRC Press. ISBN 1-56670-670-X.
- Baselt RC (2014). Disposition of toxic drugs and chemicals in man. Seal Beach, Ca.: Biomedical Publications. pp. 1213–1214. ISBN 978-0-9626523-9-4.
- Neergaard, Lauran. "Study Examines Why Most Survived China's Melamine Scare". Food Manufacturing News. Food Manufacturing. Retrieved 22 February 2013.
- "40 C.F.R.: Appendix A to Part 355—The List of Extremely Hazardous Substances and Their Threshold Planning Quantities" (PDF) (July 1, 2008 ed.). Government Printing Office. Retrieved October 29, 2011.

6

Understanding Genetically Modified Food

Genetically modified food is produced by genetically modified organisms (GMOs) to allow domination of certain genetic traits and cause recession of unwanted ones. GMOs have ensured higher nutrient value, better resistance to dry conditions, frost, pests, insects etc. as well as delayed or advanced ripening among other food products. Though GMOs have a multitude of benefits and enable scientists to produce crops that have higher yield potential, their use is still controversial as they can wipe out indigenous species of plants and their future safety for the environment and humans is still questionable.

Genetically Modified Food

Genetically modified foods or GM foods, also genetically engineered foods, are foods produced from organisms that have had changes introduced into their DNA using the methods of genetic engineering. Genetic engineering techniques allow for the introduction of new traits as well as greater control over traits than previous methods such as selective breeding and mutation breeding.

Commercial sale of genetically modified foods began in 1994, when Calgene first marketed its unsuccessful Flavr Savr delayed-ripening tomato. Most food modifications have primarily focused on cash crops in high demand by farmers such as soybean, corn, canola, and cotton. Genetically modified crops have been engineered for resistance to pathogens and herbicides and for better nutrient profiles. GM livestock have been developed, although as of November 2013 none were on the market.

There is a scientific consensus that currently available food derived from GM crops poses no greater risk to human health than conventional food, but that each GM food needs to be tested on a case-by-case basis before introduction. Nonetheless, members of the public are much less likely than scientists to perceive GM foods as safe. The legal and regulatory status of GM foods varies by country, with some nations banning or restricting them, and others permitting them with widely differing degrees of regulation.

However, there are ongoing public concerns related to food safety, regulation, labelling, environmental impact, research methods, and the fact that some GM seeds are subject to intellectual property rights owned by corporations.

Definition

Genetically modified foods, GM foods or genetically engineered foods, are foods produced from organisms that have had changes introduced into their DNA using the methods of genetic engineering as opposed to traditional cross breeding. In the US, the Department of Agriculture (USDA) and the Food and Drug Administration (FDA) favor the use of "genetic engineering" over "genetic modification" as the more precise term; the USDA defines genetic modification to include "genetic engineering or other more traditional methods."

According to the World Health Organization, "Genetically modified organisms (GMOs) can be defined as organisms (i.e. plants, animals or microorganisms) in which the genetic material (DNA) has been altered in a way that does not occur naturally by mating and/or natural recombination. The technology is often called 'modern biotechnology' or 'gene technology', sometimes also 'recombinant DNA technology' or 'genetic engineering'. ... Foods produced from or using GM organisms are often referred to as GM foods."

History

Human-directed genetic manipulation of food began with the domestication of plants and animals through artificial selection at about 10,500 to 10,100 BC. The process of selective breeding, in which organisms with desired traits (and thus with the desired genes) are used to breed the next generation and organisms lacking the trait are not bred, is a precursor to the modern concept of genetic modification (GM). With the discovery of DNA in the early 1900s and various advancements in genetic techniques through the 1970s it became possible to directly alter the DNA and genes within food.

Genetically modified microbial enzymes were the first application of genetically modified organisms in food production and were approved in 1988 by the US Food and Drug Administration. In the early 1990s, recombinant chymosin was approved for use in several countries. Cheese had typically been made using the enzyme complex rennet that had been extracted from cows' stomach lining. Scientists modified bacteria to produce chymosin, which was also able to clot milk, resulting in cheese curds.

The first genetically modified food approved for release was the Flavr Savr tomato in 1994. Developed by Calgene, it was engineered to have a longer shelf life by inserting an antisense gene that delayed ripening. In 1995, *Bacillus thuringiensis* (Bt) Potato was approved for cultivation, making it the first pesticide producing crop to be approved in the USA. Other genetically modified crops receiving marketing approval in 1995 were: canola with modified oil composition, Bt maize, cotton resistant to the herbicide bromoxynil, Bt cotton, glyphosate-tolerant soybeans, virus-resistant squash, and another delayed ripening tomato.

With the creation of golden rice in 2000, scientists had genetically modified food to increase its nutrient value for the first time.

By 2010, 29 countries had planted commercialized biotech crops and a further 31 countries had granted regulatory approval for transgenic crops to be imported. The US was the leading country in the production of GM foods in 2011, with twenty-five GM crops having received regulatory approval. In 2015, 92% of corn, 94% of soybeans, and 94% of cotton produced in the US were genetically modified strains.

The first genetically modified animal to be approved for food use was AquAdvantage salmon in 2015. The salmon were transformed with a growth hormone-regulating gene from a Pacific Chinook salmon and a promoter from an ocean pout enabling it to grow year-round instead of only during spring and summer.

In April 2016, a white button mushroom (*Agaricus bisporus*) modified using the CRISPR technique received *de facto* approval in the United States, after the USDA said it would not have to go through the agency's regulatory process. The agency considers the mushroom exempt because the editing process did not involve the introduction of foreign DNA.

The most widely planted GMOs are designed to tolerate herbicides. By 2006 some weed populations had evolved to tolerate some of the same herbicides. Palmer amaranth is a weed that competes with cotton. A native of the southwestern US, it traveled east and was first found resistant to glyphosate in 2006, less than 10 years after GM cotton was introduced.

Process

Genetically engineered organisms are generated and tested in the laboratory for desired qualities. The most common modification is to add one or more genes to an organism's genome. Less commonly, genes are removed or their expression is increased or silenced or the number of copies of a gene is increased or decreased.

Once satisfactory strains are produced, the producer applies for regulatory approval to field-test them, called a "field release." Field-testing involves cultivating the plants on farm fields or growing animals in a controlled environment. If these field tests are successful, the producer applies for regulatory approval to grow and market the crop. Once approved, specimens (seeds, cuttings, breeding pairs, etc.) are cultivated and sold to farmers. The farmers cultivate and market the new strain. In some cases, the approval covers marketing but not cultivation.

According to the USDA, the number of field releases for genetically engineered organisms has grown from four in 1985 to an average of about 800 per year. Cumulatively, more than 17,000 releases had been approved through September 2013.

Crops

Fruits and Vegetables

3 views of the Sunset papaya cultivar, which was genetically modified to create the SunUp cultivar, resistant to PRSV.

Papaya was genetically modified to resist the ringspot virus. 'SunUp' is a transgenic red-fleshed Sunset papaya cultivar that is homozygous for the coat protein gene PRSV; 'Rainbow' is a yellow-fleshed F1 hybrid developed by crossing 'SunUp' and nontransgenic yellow-fleshed 'Kapoho'. The New York Times stated, "in the early 1990s, Hawaii's papaya industry was facing disaster because of the deadly papaya ringspot virus. Its single-handed savior was a breed engineered to be resistant to the virus. Without it, the state's papaya industry would have collapsed. Today, 80% of Hawaiian papaya is genetically engineered, and there is still no conventional or organic method to control ringspot virus." The GM cultivar was approved in 1998. In China, a transgenic PRSV-resistant papaya was developed by South China Agricultural University and was first approved for commercial planting in 2006; as of 2012 95% of the papaya grown in Guangdong province and 40% of the papaya grown in Hainan province was genetically modified.

The New Leaf potato, brought to market by Monsanto in the late 1990s, was developed for the fast food market. It was withdrawn in 2001 after retailers rejected it and food processors ran into export problems.

As of 2005, about 13% of the Zucchini (a form of squash) grown in the US was genetically modified to resist three viruses; that strain is also grown in Canada.

Plums genetically engineered for resistance to plum pox, a disease carried by aphids.

In 2011, BASF requested the European Food Safety Authority's approval for cultivation and marketing of its Fortuna potato as feed and food. The potato was made resistant to late blight by adding resistant genes blb1 and blb2 that originate from the Mexican wild potato Solanum bulbocastanum. In February 2013, BASF withdrew its application.

In 2013, the USDA approved the import of a GM pineapple that is pink in color and that "overexpresses" a gene derived from tangerines and suppress other genes, increasing production of lycopene. The plant's flowering cycle was changed to provide for more uniform growth and quality. The fruit "does not have the ability to propagate and persist in the environment once they have been harvested," according to USDA APHIS. According to Del Monte's submission, the pineapples are commercially grown in a "monoculture" that prevents seed production, as the plant's flowers aren't exposed to compatible pollen sources. Importation into Hawaii is banned for "plant sanitation" reasons.

In 2014, the USDA approved a genetically modified potato developed by J.R. Simplot Company that contained ten genetic modifications that prevent bruising and produce less acrylamide when fried. The modifications eliminate specific proteins from the potatoes, via RNA interference, rather than introducing novel proteins.

In February 2015 Arctic Apples were approved by the USDA, becoming the first genetically modified apple approved for sale in the US. Gene silencing is used to reduce the expression of polyphenol oxidase (PPO), thus preventing the fruit from browning.

Corn

Corn used for food and ethanol has been genetically modified to tolerate various herbicides and to express a protein from Bacillus thuringiensis (Bt) that kills certain insects. About 90% of the corn grown in the U.S. was genetically modified in 2010. In the US in 2015, 81% of corn acreage contained the Bt trait and 89% of corn acreage contained the glyphosate-tolerant trait. Corn can be processed into grits, meal and flour as an ingredient in pancakes, muffins, doughnuts, breadings and batters, as well as baby foods, meat products, cereals and some fermented products. Corn-based masa flour and masa dough are used in the production of taco shells, corn chips and tortillas.

Soy

Genetically modified soybean has been modified to tolerate herbicides and produce healthier oils. In 2015, 94% of soybean acreage in the U.S. was genetically modified to be glyphosate-tolerant.

Derivative Products

Corn Starch and starch Sugars, Including Syrups

Starch or amylum is a polysaccharide produced by all green plants as an energy store. Pure starch is a white, tasteless and odourless powder. It consists of two types of molecules: the linear and helical amylose and the branched amylopectin. Depending on the plant, starch generally contains 20 to 25% amylose and 75 to 80% amylopectin by weight.

Starch can be further modified to create modified starch for specific purposes, including creation of many of the sugars in processed foods. They include:

- Maltodextrin, a lightly hydrolyzed starch product used as a bland-tasting filler and thickener.
- Various glucose syrups, also called corn syrups in the US, viscous solutions used as sweeteners and thickeners in many kinds of processed foods.
- Dextrose, commercial glucose, prepared by the complete hydrolysis of starch.
- High fructose syrup, made by treating dextrose solutions with the enzyme glucose isomerase, until a substantial fraction of the glucose has been converted to fructose.
- Sugar alcohols, such as maltitol, erythritol, sorbitol, mannitol and hydrogenated starch hydrolysate, are sweeteners made by reducing sugars.

Lecithin

Lecithin is a naturally occurring lipid. It can be found in egg yolks and oil-producing plants. it is an emulsifier and thus is used in many foods. Corn, soy and safflower oil are sources of lecithin, though the majority of lecithin commercially available is derived from soy.Sufficiently processed lecithin is often undetectable with standard testing practices.According to the FDA, no evidence shows or suggests hazard to the public when lecithin is used at common levels. Lecithin added to foods amounts to only 2 to 10 percent of the 1 to 5 g of phosphoglycerides consumed daily on average. Nonetheless, consumer concerns about GM food extend to such products.This concern led to policy and regulatory changes in Europe in 2000, when Regulation (EC) 50/2000 was passed which required labelling of food containing additives derived from GMOs, including lecithin. Because of the difficulty of detecting the origin of derivatives like lecithin with current testing practices, European regulations require those who wish to sell lecithin in Europe to employ a comprehensive system of Identity preservation (IP).

Sugar

The US imports 10% of its sugar, while the remaining 90% is extracted from sugar beet and sugarcane. After deregulation in 2005, glyphosate-resistant sugar beet was extensively adopted in the United States. 95% of beet acres in the US were planted with glyphosate-resistant seed in 2011. GM sugar beets are approved for cultivation in the US, Canada and Japan; the vast majority are grown in the US. GM beets are approved for import and consumption in Australia, Canada, Colombia, EU, Japan, Korea, Mexico, New Zealand, Philippines, Russian Federation and Singapore. Pulp from the refining process is used as animal feed. The sugar produced from GM sugarbeets contains no DNA or protein—it is just sucrose that is chemically indistinguishable from sugar produced from non-GM sugarbeets. Independent analyses conducted by internationally recognized laboratories found that sugar from Roundup Ready sugar beets is identical to the sugar from comparably grown conventional (non-Roundup Ready) sugar beets. And, like all sugar, sugar from Roundup Ready sugar beets contains no genetic material or detectable protein (including the protein that provides glyphosate tolerance).

Vegetable Oil

Most vegetable oil used in the US is produced from GM crops canola, corn, cotton and soybeans. Vegetable oil is sold directly to consumers as cooking oil, shortening and margarine and is used in prepared foods. There is a vanishingly small amount of protein or DNA from the original crop in vegetable oil. Vegetable oil is made of triglycerides extracted from plants or seeds and then refined and may be further processed via hydrogenation to turn liquid oils into solids. The refining process removes all, or nearly all non-triglyceride ingredients.

Other Uses

Animal feed

Livestock and poultry are raised on animal feed, much of which is composed of the leftovers from processing crops, including GM crops. For example, approximately 43% of a canola seed is oil. What remains after oil extraction is a meal that becomes an ingredient in animal feed and contains canola protein. Likewise, the bulk of the soybean crop is grown for oil and meal. The high-protein defatted and toasted soy meal becomes livestock feed and dog food. 98% of the US soybean crop goes for livestock feed. In 2011, 49% of the US maize harvest was used for livestock feed (including the percentage of waste from distillers grains). "Despite methods that are becoming more and more sensitive, tests have not yet been able to establish a difference in the meat, milk, or eggs of animals depending on the type of feed they are fed. It is impossible to tell if an animal was fed GM soy just by looking at the resulting meat, dairy, or egg products. The only way to verify the presence of GMOs in animal feed is to analyze the origin of the feed itself."

A 2012 literature review of studies evaluating the effect of GM feed on the health of animals did not find evidence that animals were adversely affected, although small biological differences were occasionally found. The studies included in the review ranged from 90 days to two years, with several of the longer studies considering reproductive and intergenerational effects.

Proteins

Rennet is a mixture of enzymes used to coagulate milk into cheese. Originally it was available only from the fourth stomach of calves, and was scarce and expensive, or was available from microbial sources, which often produced unpleasant tastes. Genetic engineering made it possible to extract rennet-producing genes from animal stomachs and insert them into bacteria, fungi or yeasts to make them produce chymosin, the key enzyme. The modified microorganism is killed after fermentation. Chymosin is isolated from the fermentation broth, so that the Fermentation-Produced Chymosin (FPC) used by cheese producers has an amino acid sequence that is identical to bovine rennet. The majority of the applied chymosin is retained in the whey. Trace quantities of chymosin may remain in cheese.

FPC was the first artificially produced enzyme to be approved by the US Food and Drug Administration. FPC products have been on the market since 1990 and as of 2015 had yet to be surpassed in commercial markets. In 1999, about 60% of US hard cheese was made with FPC. Its global market share approached 80%. By 2008, approximately 80% to 90% of commercially made cheeses in the US and Britain were made using FPC. The most widely used FPC is produced either by the fungus *Aspergillus niger* (CHY-MAX®)

In some countries, recombinant (GM) bovine somatotropin (also called rBST, or bovine growth hormone or BGH) is approved for administration to increase milk production. rBST may be present in milk from rBST treated cows, but it is destroyed in the digestive system and even if directly injected into the human bloodstream, has no observable effect on humans. The FDA, World Health Organization, American Medical Association, American Dietetic Association and the National Institutes of Health have independently stated that dairy products and meat from rBST-treated cows are safe for human consumption. However, on 30 September 2010, the United States Court of Appeals, Sixth Circuit, analyzing submitted evidence, found a "compositional difference" between milk from rBGH-treated cows and milk from untreated cows. The court stated that milk from rBGH-treated cows has: increased levels of the hormone Insulin-like growth factor 1 (IGF-1); higher fat content and lower protein content when produced at certain points in the cow's lactation cycle; and more somatic cell counts, which may "make the milk turn sour more quickly."

Livestock

Genetically modified livestock are organisms from the group of cattle, sheep, pigs, goats, birds, horses and fish kept for human consumption, whose genetic material (DNA) has been altered using genetic engineering techniques. In some cases, the aim is to introduce a new trait to the animals which does not occur naturally in the species, i.e. transgenesis.

A 2003 review published on behalf of Food Standards Australia New Zealand examined transgenic experimentation on terrestrial livestock species as well as aquatic species such as fish and shellfish. The review examined the molecular techniques used for experimentation as well as techniques for tracing the transgenes in animals and products as well as issues regarding transgene stability.

Some mammals typically used for food production have been modified to produce non-food products, a practice sometimes called Pharming.

Salmon

A GM salmon, awaiting regulatory approval since 1997, was approved for human consumption by the American FDA in November 2015, to be raised in specific land-based hatcheries in Canada and Panama.

Recombinant Food-Grade Organisms for Healthcare

The use of genetically modified food-grade organisms as recombinant vaccine expression hosts and delivery vehicles can open new avenues for vaccinology. Considering that oral immunization is a beneficial approach in terms of costs, patient comfort, and protection of mucosal tissues, the use of food-grade organisms can lead to highly advantageous vaccines in terms of costs, easy administration, and safety. The organisms currently used for this purpose are bacteria (Lactobacillus and Bacillus), yeasts, algae, plants, and insect species. Several such organisms are under clinical evaluation, and the current adoption of this technology by the industry indicates a potential to benefit global healthcare systems.

Health and Safety

There is a scientific consensus that currently available food derived from GM crops poses no greater risk to human health than conventional food, but that each GM food needs to be tested on a case-by-case basis before introduction. Nonetheless, members of the public are much less likely than scientists to perceive GM foods as safe.

Opponents claim that long-term health risks have not been adequately assessed and propose various combinations of additional testing, labeling or removal from the market. The advocacy group European Network of Scientists for Social and Environmental Responsibility (ENSSER), disputes the claim that "science" supports the safety of current GM foods, proposing that each GM food must be judged on case-by-case basis. The Canadian Association of Physicians for the Environment called for removing GM foods from the market pending long term health studies. Multiple disputed studies have claimed health effects relating to GM foods or to the pesticides used with them.

Testing

The legal and regulatory status of GM foods varies by country, with some nations banning or restricting them, and others permitting them with widely differing degrees of regulation. Countries such as the United States, Canada, Lebanon and Egypt use *substantial equivalence* to determine if further testing is required, while many countries such as those in the European Union, Brazil and China only authorize GMO cultivation on a case-by-case basis. In the U.S. the FDA determined that GMO's are "Generally Recognized as Safe" (GRAS) and therefore do not require additional testing if the GMO product is substantially equivalent to the non-modified product. If new substances are found, further testing may be required to satisfy concerns over potential toxicity, allergenicity, possible gene transfer to humans or genetic outcrossing to other organisms.

Regulation

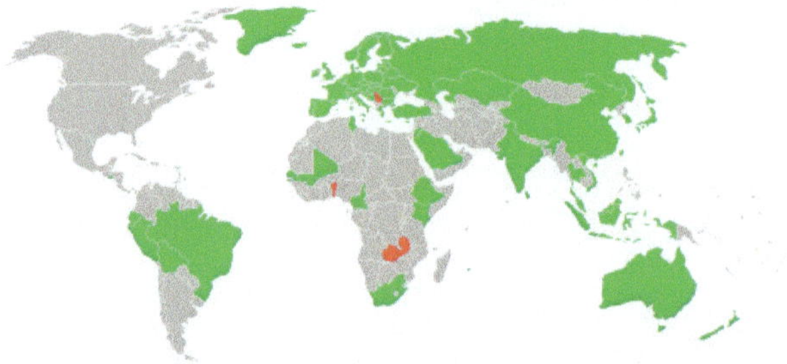

Green: Mandatory labeling required; Red:Ban on import and cultivation of genetically engineered food.

Government regulation of GMO development and release varies widely between countries. Marked differences separate GMO regulation in the U.S. and GMO regulation in the European Union. Regulation also varies depending on the intended product's use. For example, a crop not intended for food use is generally not reviewed by authorities responsible for food safety.

United States Regulations

In the U.S., three government organizations regulate GMOs. The FDA checks the chemical composition of organisms for potential allergens. The United States Department of Agriculture (USDA) supervises field testing and monitors the distribution of GM seeds. The United States Environmental Protection Agency (EPA) is responsible for monitoring pesticide usage, including plants modified to contain proteins toxic to insects. Like USDA, EPA also oversees field testing and the distribution of crops that have had contact with pesticides to ensure environmental safety. In 2015 the Obama administration announced that it would update the way the government regulated GM crops.

In 1992 FDA published "Statement of Policy: Foods derived from New Plant Varieties." This statement is a clarification of FDA's interpretation of the Food, Drug, and Cosmetic Act with respect to foods produced from new plant varieties developed using recombinant deoxyribonucleic acid (rDNA) technology. FDA encouraged developers to consult with the FDA regarding any bioengineered foods in development. The FDA says developers routinely do reach out for consultations. In 1996 FDA updated consultation procedures.

Labeling

As of 2015, 64 countries require labeling of GMO products in the marketplace.

US and Canadian national policy is to require a label only given significant composition differences or documented health impacts, although some individual US states (Vermont, Connecticut and Maine) enacted laws requiring them.

In some jurisdictions, the labeling requirement depends on the relative quantity of GMO in the product. A study that investigated voluntary labeling in South Africa found that 31% of products labeled as GMO-free had a GM content above 1.0%.

In Europe all food (including processed food) or feed that contains greater than 0.9% GMOs must be labelled.

Detection

Testing on GMOs in food and feed is routinely done using molecular techniques such as PCR and bioinformatics.

In a January 2010 paper, the extraction and detection of DNA along a complete industrial soybean oil processing chain was described to monitor the presence of Roundup Ready (RR) soybean: "The amplification of soybean lectin gene by end-point polymerase chain reaction (PCR) was successfully achieved in all the steps of extraction and refining processes, until the fully refined soybean oil. The amplification of RR soybean by PCR assays using event-specific primers was also achieved for all the extraction and refining steps, except for the intermediate steps of refining (neutralisation, washing and bleaching) possibly due to sample instability. The real-time PCR assays using specific probes confirmed all the results and proved that it is possible to detect and quantify genetically modified organisms in the fully refined soybean oil. To our knowledge, this has never been reported before and represents an important accomplishment regarding the traceability of genetically modified organisms in refined oils."

According to Thomas Redick, detection and prevention of cross-pollination is possible through the suggestions offered by the Farm Service Agency (FSA) and Natural Resources Conservation Service (NRCS). Suggestions include educating farmers on the importance of coexistence, providing farmers with tools and incentives to promote coexistence, conduct research to understand and monitor gene flow, provide assurance of quality and diversity in crops, provide compensation for actual economic losses for farmers.

Controversies

The genetically modified foods controversy consists of a set of disputes over the use of food made from genetically modified crops. The disputes involve consumers, farmers, biotechnology companies, governmental regulators, non-governmental organizations, environmental and political activists and scientists. The major disagreements include whether GM foods can be safely consumed, harm the environment and/or are adequately tested and regulated. The objectivity of scientific research and publications has been challenged. Farming-related disputes include the use and impact of pesticides, seed production and use, side effects on non-GMO crops/farms, and potential control of the GM food supply by seed companies.

The conflicts have continued since GM foods were invented. They have occupied the media, the courts, local, regional and national governments and international organizations.

Genetically Modified Food Controversies

Genetically modified food controversies are disputes over the use of foods and other goods derived from genetically modified crops instead of conventional crops, and other uses of genetic engineering in food production. The dispute involves consumers, farmers, biotechnology companies, governmental regulators, non-governmental organizations, and scientists. The key areas of controversy related to genetically modified food (GM food or GMO food) are whether such food should be labeled, the role of government regulators, the objectivity of scientific research and publication, the effect of genetically modified crops on health and the environment, the effect on pesticide resistance, the impact of such crops for farmers, and the role of the crops in feeding the world population. In addition, products derived from GMO organisms play a role in the production of ethanol fuels and pharmaceuticals.

Specific concerns include mixing of genetically modified and non-genetically modified products in the food supply, effects of GMOs on the environment, the rigor of the regulatory process, and consolidation of control of the food supply in companies that make and sell GMOs. Advocacy groups such as the Center for Food Safety, Organic Consumers Association, Union of Concerned Scientists, and Greenpeace, say risks have not been adequately identified and managed, and they have questioned the objectivity of regulatory authorities.

The safety assessment of genetically engineered food products by regulatory bodies starts with an evaluation of whether or not the food is *substantially equivalent* to non-genetically engineered counterparts that are already deemed fit for human consumption. No reports of ill effects have been documented in the human population from genetically modified food. There is a scientific consensus that currently available food derived from GM crops poses no greater risk to human health than conventional food, but that each GM food needs to be tested on a case-by-case basis before introduction. Nonetheless, members of the public are much less likely than scientists to perceive GM foods as safe. The legal and regulatory status of GM foods varies by country, with some nations banning or restricting them, and others permitting them with widely differing degrees of regulation.

Public Perception

Consumer concerns about food quality first became prominent long before the advent of GM foods in the 1990s. Upton Sinclair's novel *The Jungle* led to the 1906 Pure Food and Drug Act, the first major US legislation on the subject. This began an enduring concern over the purity and later "naturalness" of food that evolved from a focus on sanitation to include added ingredients such as preservatives and flavors and sweeteners, residues such as pesticides, the rise of organic food as a category and finally to concerns over GM food. The public came to see the latter as "unnatural" which created a reverse halo effect.

Specific perceptions include genetic engineering as meddling with naturally evolved biological processes, scientific limitations on comprehending potential negative ramifications. An opposing perception is that genetic engineering is itself an evolution of traditional selective breeding.

Surveys indicate public concerns that eating genetically modified food is harmful, that biotechnology is risky, that more information is needed and that consumers need control over whether to take such risks. A diffuse sense that social and technological change is accelerating and that people cannot affect this change context becomes focused when such changes affect food. Leaders in driving public perception of the harms of such food in the media include Jeffrey M. Smith, Dr. Oz, Oprah, and Bill Maher; organizations include Organic Consumers Association, Greenpeace (especially with regard to Golden rice) and Union of Concerned Scientists.

Religious groups have raised concerns over whether genetically modified food will remain kosher or halal. In 2001 no such foods had been designated as unacceptable by Orthodox rabbis or Muslim leaders. However, some Jewish groups dispute this designation.

Environmental groups such as Friends of the Earth, include genetic engineering in general as an environmental and political concern. Other groups such as GMWatch and The Institute of Science in Society concentrate mostly or solely on opposing genetically modified crops.

Food writer Michael Pollan does not oppose eating genetically modified foods, but expressed concerns about biotechnology companies holding the intellectual property of the foods people depend on, and about the effects of the growing corporatization of large-scale agriculture. To address these problems, Pollan has brought up the idea of open sourcing GM foods. The idea has since been adopted to varying degrees by companies like Syngenta, and is being promoted by organizations such as the New America Foundation. Some organizations, like The BioBricks Foundation, have already worked out open-source licenses that could prove useful in this endeavour.

Reviews and Polls

A 2003 *EMBO Reports* article reported that the *Public Perceptions of Agricultural Biotechnologies in Europe* project (PABE) found the public neither accepting nor rejecting GMOs. Instead, PABE found that public had "key questions" about GMOs: "Why do we need GMOs? Who benefits from their use? Who decided that they should be developed and how? Why were we not better informed about their use in our food, before their arrival on the market? Why are we not given an effective choice about whether or not to buy these products? Have potential long-term and irreversible consequences been seriously evaluated, and by whom? Do regulatory authorities have sufficient powers to effectively regulate large companies? Who wishes to develop these products? Can controls imposed by regulatory authorities be applied effectively? Who will be accountable in cases of unforeseen harm?" PABE also found that the public's scientific knowledge does not control public opinion, since scientific facts do not answer these questions. PABE also found that the public does not demand "zero risk" in GM food discussions and is "perfectly aware that their lives are full of risks that need to be counterbalanced against each other and against the potential benefits. Rather than zero risk, what they demanded was a more realistic assessment of risks by regulatory authorities and GMO producers."

In 2006, the Pew Initiative on Food and Biotechnology made public a review of U.S. survey results from 2001-2006. The review showed that Americans' knowledge of GM foods and animals was low throughout the period. During this period protests against Calgene's Flavr Savr GM tomato mistakenly described it as containing fish genes, confusing it with DNA Plant Technology's fish tomato experimental transgenic organism, which was never commercialized.

A 2007 survey by the Food Standards Australia New Zealand found that in Australia, where labeling is mandatory, 27% of Australians checked product labels to see whether GM ingredients were present when initially purchasing a food item.

A 2009 review article of European consumer polls concluded that opposition to GMOs in Europe has been gradually decreasing, and that about 80% of respondents did not "actively avoid GM products when shopping". The 2010 "Eurobarometer" survey, which assesses public attitudes about biotech and the life sciences, found that cisgenics, GM crops made from plants that are crossable by conventional breeding, evokes a smaller reaction than transgenic methods, using genes from species that are taxonomically very different.

A 2010 Deloitte survey found that 34% of U.S. consumers were very or extremely concerned about GM food, a 3% reduction from 2008. The same survey found gender differences: 10% of men were extremely concerned, compared with 16% of women, and 16% of women were unconcerned, compared with 27% of men.

A 2013 poll by *The New York Times* showed that 93% of Americans wanted labeling of GM food.

The 2013 vote rejecting Washington State's GM food labeling I-522 referendum came shortly after the 2013 World Food Prize was awarded to employees of Monsanto and Syngenta. The award has drawn criticism from opponents of genetically modified crops.

With respect to the question of "Whether GMO foods were safe to eat," the gap between the opinion of the public and that of American Association for the Advancement of Science scientists is very wide with 88% of AAAS scientists saying yes in contrast to 37% of the general public.

Public Relations Campaigns and Protests

Anti-GMO and anti-Monsanto protests in Washington, D.C.

March Against Monsanto in Stockholm, Sweden, May 2013

In May 2012, a group called "Take the Flour Back" led by Gerald Miles protested plans by a group from Rothamsted Experimental Station, based in Harpenden, Hertfordshire, England, to conduct an experimental trial wheat genetically modified to repel aphids. The researchers, led by John Pickett, wrote a letter to the group in early May 2012, asking them to call off their protest, aimed for 27 May 2012. Group member Lucy Harrap said that the group was concerned about spread of the crops into nature, and cited examples of outcomes in the United States and Canada. Rothamsted Research and Sense About Science ran question and answer sessions about such a potential.

The March Against Monsanto is an international grassroots movement and protest against Monsanto corporation, a producer of genetically modified organism (GMOs) and Roundup, a glyphosate-based herbicide. The movement was founded by Tami Canal in response to the failure of California Proposition 37, a ballot initiative which would have required labeling food products made from GMOs. Advocates support mandatory labeling laws for food made from GMOs .

The initial march took place on May 25, 2013. The number of protesters who took part is uncertain; figures of "hundreds of thousands" and the organizers' estimate of "two million" were variously cited. Events took place in between 330 and 436 cities around the world, mostly in the United States. Many protests occurred in Southern California, and some participants carried signs expressing support for mandatory labeling of GMOs that read "Label GMOs, It's Our Right to Know", and "Real Food 4 Real People". Canal said that the movement would continue its "anti-GMO cause" beyond the initial event. Further marches occurred in October 2013 and in May 2014 and 2015. The protests were reported by news outlets including ABC News, the Associated Press, *The Washington Post*, *The Los Angeles Times*, *USA Today*, and CNN (in the United States), and *Russia Today* and *The Guardian* (outside the United States).

Monsanto said that it respected people's rights to express their opinion on the topic, but maintained that its seeds improved agriculture by helping farmers produce more from their land while conserving resources, such as water and energy. The company reiterated that genetically modified foods were safe and improved crop yields. Similar sentiments were expressed by the Hawaii Crop Improvement Association, of which Monsanto is a member.

In July 2013, the agricultural biotechnology industry launched a GMO transparency initiative called GMO Answers to address consumers' questions about GM foods in the U.S. food supply. GMO Answers' resources included conventional and organic farmers, agribusiness experts, scientists, academics, medical doctors and nutritionists, and "company experts" from founding members of the Council for Biotechnology Information, which funds the initiative. Founding members include BASF, Bayer CropScience, Dow AgroSciences, DuPont, Monsanto Company and Syngenta.

In October 2013, a group called The European Network of Scientists for Social and Environmental Responsibility (ENSSER), posted a statement claiming that there is no scientific consensus on the safety of GMOs, which was signed by about 200 scientists in various fields in its first week. On January 25, 2015, their statement was formally published as a whitepaper by Environmental Sciences Europe:

Direct Action

Earth Liberation Front, Greenpeace and others have disrupted GMO research around the world. Within the UK and other European countries, as of 2014 80 crop trials by academic or governmental research institutes had been destroyed by protesters. In some cases, threats and violence against people or property were carried out. In 1999, activists burned the biotech lab of Michigan State University, destroying the results of years of work and property worth $400,000.

In 1987, the ice-minus strain of *P. syringae* became the first genetically modified organism (GMO) to be released into the environment when a strawberry field in California was sprayed with the bacteria. This was followed by the spraying of a crop of potato seedlings. The plants in both test fields were uprooted by activist groups, but were re-planted the next day.

In 2011, Greenpeace paid reparations when its members broke into the premises of an Australian scientific research organization, CSIRO, and destroyed a genetically modified wheat plot. The sentencing judge accused Greenpeace of cynically using junior members to avoid risking their own freedom. The offenders were given 9-month suspended sentences.

On August 8, 2013 protesters uprooted an experimental plot of golden rice in the Philippines. British author, journalist, and environmental activist Mark Lynas reported in *Slate* that the vandalism was carried out by a group led by the extreme-left KMP, to the dismay of other protesters. Golden rice is designed prevent vitamin A deficiency which, according to Helen Keller International, blinds or kills hundreds of thousands of children annually in developing countries.

Conspiracy Theories

There are various conspiracy theories related to the production and sale of genetically modified crops and genetically modified food that have been identified by some commentators such as Michael Shermer. Generally, these conspiracy theories posit that GMOs are being knowingly and maliciously introduced into the food supply either as a means to unduly enrich agribusinesses or as a means to poison or pacify the population.

A work seeking to explore risk perception over GMOs in Turkey identified a belief among the conservative political and religious figures who were opposed to GMOs that GMOs were "a conspiracy by Jewish Multinational Companies and Israel for world domination." Additionally, a Latvian study showed that a segment of the population believed that GMOs were part of a greater conspiracy theory to poison the population of the country.

Lawsuits

Foundation on Economic Trends v. Heckler

In 1983, environmental groups and protesters delayed the field tests of the genetically modified ice-minus strain of *P. syringae* with legal challenges.

Alliance for Bio-Integrity v. Shalala

In this case, the plaintiff argued both for mandatory labeling on the basis of consumer demand, and that GMO foods should undergo the same testing requirements as food additives because they are "materially changed" and have potentially unidentified health risks. The plaintiff also alleged that the FDA did not follow the Administrative Procedures Act in formulating and disseminating its policy on GMO's. The federal district court rejected all of those arguments and found that the FDA's determination that GMO's are Generally Recognized as Safe was neither arbitrary nor capricious. The court gave deference to the FDA's process on all issues, leaving future plaintiffs little legal recourse to challenge the FDA's policy on GMO's.

Diamond v. Chakrabarty

This case was on the question of whether GMOs can be patented.

Scientific Publishing

Scientific publishing on the safety and effects of GM foods is controversial.

Bt Maize

One of the first incidents occurred in 1999, when *Nature* published a paper on potential toxic effects of Bt maize on butterflies. The paper produced a public uproar and demonstrations, however by 2001 multiple follow-up studies had concluded that "the most common types of Bt maize pollen are not toxic to monarch larvae in concentrations the insects would encounter in the fields" and that they had "brought that particular question to a close."

Concerned scientists began to patrol the scientific literature and react strongly, both publicly and privately, to discredit conclusions they view as flawed in order to prevent unjustified public outcry and regulatory action. A 2013 *Scientific American* article noted that a "tiny minority" of biologists have published concerns about GM food, and said that scientists who support the use of GMOs in food production are often overly dismissive of them.

Restrictive End-User Agreements

Prior to 2010, scientists wishing to conduct research on commercial GM plants or seeds were unable to do so, because of restrictive end-user agreements. Cornell University's Elson Shields was the spokesperson for one group of scientists who opposed such restrictions. The group submitted a statement to the United States Environmental Protection Agency (EPA) in 2009 protesting that "as a result of restrictive access, no truly independent research can be legally conducted on many critical questions regarding the technology".

A 2009 *Scientific American* editorial quoted a scientist who said that several studies that were initially approved by seed companies were blocked from publication when they returned "unflattering" results. While favoring protection of intellectual property rights, the editors called for the restrictions to be lifted and for the EPA to require, as a condition of approval, that independent researchers have unfettered access to genetically modified products for research.

In December 2009, the American Seed Trade Association agreed to "allow public researchers greater freedom to study the effects of GM food crops." The companies signed blanket agreements permitting such research. This agreement left many scientists optimistic about the future, However, some scientists still express concern as to whether this agreement has the ability to "alter what has been a research environment rife with obstruction and suspicion." Monsanto previously had research agreements (i.e., Academic Research Licenses) with approximately 100 universities that allowed for university scientists to conduct research on their GM products with no oversight.

Reviews

A 2011 analysis by Diels *et al.* reviewed 94 peer-reviewed studies pertaining to GMO safety to assess whether conflicts of interest correlated with outcomes that cast GMOs in a favorable light. They found that financial conflict of interest was not associated with study outcome ($p = 0.631$) while author affiliation to industry (i.e., a professional conflict of interest) was strongly associated with study outcome ($p < 0.001$). Of the 94 studies that were analyzed, 52% did not declare funding. 10% of the studies were categorized as "undetermined" with regard to professional conflict of interest. Of the 43 studies with financial or professional conflicts of interest, 28 studies were compositional studies. According to Marc Brazeau, an association between professional conflict of interest and positive study outcomes can be skewed because companies typically contract with independent researchers to perform follow-up studies only after in-house research uncovers favorable results. In-house research that uncovers negative or unfavorable results for a novel GMO is generally not further pursued.

A 2013 review of 1,783 papers on genetically modified crops and food published between 2002 and 2012 found no plausible evidence of dangers from the use of then marketed GM crops. Biofortified, an independent nonprofit organization devoted to providing factual information and fostering discussion about agriculture, especially plant genetics and genetic engineering, planned to add the studies found by the Italian group to its database of studies about GM crops, GENERA.

In a 2014 review, Zdziarski *et al.* examined published studies of the histopathology of GI tracts of rats that were fed diets derived from GM crops, and identified some systemic flaws in this area of the scientific literature. Most studies were performed years after the approval of the crop for human consumption. Papers were often imprecise in their descriptions of the histological results and the selection of study endpoints, and lacked necessary details about methods and results. The authors called for the development of better study guidelines for determining the long-term safety of eating GM foods.

A 2016 study by the US National Academies of Sciences, Engineering, and Medicine concluded that GM foods are safe for human consumption and they could find no conclusive evidence that they harm the environment nor wildlife. They analysed over 1.000 studies over the previous 30 years that GM crops have been available, reviewed 700 written presentations submitted by interested bodies and heard 80 witnesses. They concluded that GM crops had given farmers economic advantages but found no evidence that GM crops had increased yields. They also noted that weed resistance to GM crops could cause major agricultural problems but this could be addressed by better farming procedures.

Alleged Data Manipulation

A University of Naples investigation suggested that images in eight papers on animals were intentionally altered and/or misused. The leader of the research group, Federico Infascelli, rejected the claim. The research concluded that mother goats fed GM soybean meal secreted fragments of the foreign gene in their milk. In December 2015 one of the papers was retracted for "self-plagiarism", although the journal noted that the results remained valid. A second paper was retracted in March 2016 after The University of Naples concluded that "multiple heterogeneities were likely attributable to digital manipulation, raising serious doubts on the reliability of the findings".

Health

There is a scientific consensus that currently available food derived from GM crops poses no greater risk to human health than conventional food, but that each GM food needs to be tested on a case-by-case basis before introduction. Nonetheless, members of the public are much less likely than scientists to perceive GM foods as safe. The legal and regulatory status of GM foods varies by country, with some nations banning or restricting them, and others permitting them with widely differing degrees of regulation.

The ENTRANSFOOD project was a European Commission-funded scientist group chartered to set a research program to address public concerns about the safety and value of agricultural biotechnology. It concluded that "the combination of existing test methods provides a sound test-regime to assess the safety of GM crops." In 2010, the European Commission Directorate-General for Research and Innovation reported that "The main conclusion to be drawn from the efforts of more than 130 research projects, covering a period of more than 25 years of involving more than 500 independent research groups, is that biotechnology, and in particular GMOs, are not per se more risky than e.g. conventional plant breeding technologies."

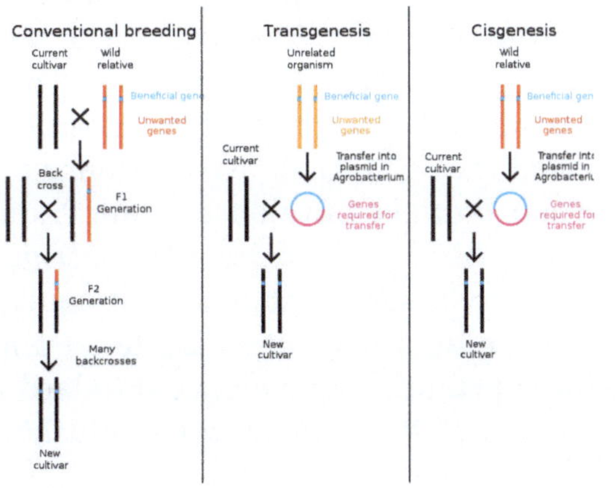

Comparison of conventional plant breeding with transgenic and cisgenic genetic modification.

Consensus among scientists and regulators pointed to the need for improved testing technologies and protocols. Transgenic and cisgenic organisms are treated similarly when assessed. However, in 2012 the European Food Safety Authority (EFSA) GMO Panel said that "novel hazards" could be associated with transgenic strains. In a 2016 review, Domingo concluded that studies in recent years had established that GM soybeans, rice, corn, and wheat do not differ from the corresponding conventional crops in terms of short-term human health effects, but recommended that further studies of long-term effects be conducted.

Substantial Equivalence

Most conventional agricultural products are the products of genetic manipulation via traditional cross-breeding and hybridization.

Governments manage the marketing and release of GM foods on a case-by-case basis. Countries differ in their risk assessments and regulations. Marked differences distinguish the US from Europe. Crops not intended as foods are generally not reviewed for food safety. GM foods are not tested in humans before marketing because they are not a single chemical, nor are they intended to be ingested using specific doses and intervals, which complicate clinical study design. Regulators examine the genetic modification, related protein products and any changes that those proteins make to the food.

Regulators check that GM foods are "substantially equivalent" to their conventional counterparts, to detect any negative unintended consequences. New protein(s) that differ from conventional food proteins or anomalies that arise in the substantial equivalence comparison require further toxicological analysis.

"The World Health Organization, the American Medical Association, the U.S. National Academy of Sciences, the British Royal Society, and every other respected organization that has examined the evidence has come to the same conclusion: consuming foods containing ingredients derived from GM crops is no riskier than consuming the same foods containing ingredients from crop plants modified by conventional plant improvement techniques."

–American Association for the Advancement of Science

In 1999, Andrew Chesson of the Rowett Research Institute warned that substantial equivalence testing "could be flawed in some cases" and that current safety tests could allow harmful substances to enter the human food supply. The same year Millstone, Brunner and Mayer argued that the standard was a pseudo-scientific product of politics and lobbying that was created to reassure consumers and aid biotechnology companies to reduce the time and cost of safety testing. They suggested that GM foods have extensive biological, toxicological and immunological tests and that substantial equivalence should be abandoned. This commentary was criticized for misrepresenting history, for distorting existing data and poor logic. Kuiper claimed that it oversimplified safety assessments and that equivalence testing involves more than chemical tests, possibly including toxicity testing. Keler and Lappe supported Congressional legislation to replace the substantial equivalence standard with safety studies. In a 2016 review, Domingo criticized the use of the "substantial equivalence" concept as a measure of the safety of GM crops.

Kuiper examined this process further in 2002, finding that substantial equivalence does not measure absolute risks, but instead identifies differences between new and existing products. He claimed that characterizing differences is properly a starting point for a safety assessment and "the concept of substantial equivalence is an adequate tool in order to identify safety issues related to genetically modified products that have a traditional counterpart". Kuiper noted practical difficulties in applying this standard, including the fact that traditional foods contain many toxic or carcinogenic chemicals and that existing diets were never proven to be safe. This lack of knowledge re conventional food means that modified foods may differ in anti-nutrients and natural toxins that have never been identified in the original plant, possibly allowing harmful changes to be missed. In turn, positive modifications may also be missed. For example, corn damaged by insects often contains high levels of fumonisins, carcinogenic toxins made by fungi that travel on insects' backs and that grow in the wounds of damaged corn. Studies show that most Bt corn has lower levels of fumonisins than conventional insect-damaged corn. Workshops and consultations organized by the OECD, WHO, and FAO have worked to acquire data and develop better understanding of conventional foods, for use in assessing GM foods.

A survey of publications comparing the intrinsic qualities of modified and conventional crop lines (examining genomes, proteomes and metabolomes) concluded that GM crops had less impact on gene expression or on protein and metabolite levels than the variability generated by conventional breeding.

In a 2013 review Herman (Dow AgroSciences) and Price (FDA, retired) argued that transgenesis is less disruptive than traditional breeding techniques because the latter routinely involve more changes (mutations, deletions, insertions and rearrangements) than the relatively limited changes (often single gene) in genetic engineering. The FDA found that all of the 148 transgenic events that they evaluated to be substantially equivalent to their conventional counterparts, as have Japanese regulators for 189 submissions including combined-trait products. This equivalence was confirmed by more than 80 peer-reviewed publications. Hence, the authors argue, compositional equivalence studies uniquely required for GM food crops may no longer be justified on the basis of scientific uncertainty.

Allergenicity

A well-known risk of genetic modification is the introduction of an allergen. Allergen testing is routine for products intended for food, and passing those tests is part of the regulatory requirements. Organizations such as the European Green Party and Greenpeace emphasize this risk. A 2005 review of the results from allergen testing stated that "no biotech proteins in foods have been documented to cause allergic reactions". Regulatory authorities require that new modified foods be tested for allergenicity before they are marketed.

GMO proponents note that because of the safety testing requirements, the risk of introducing a plant variety with a new allergen or toxin is much smaller than from traditional breeding processes, which do not require such tests. Genetic engineering can have less impact on the expression of genomes or on protein and metabolite levels than conventional breeding or (non-directed) plant mutagenesis. Toxicologists note that "conventional food is not risk-free; allergies occur with many known and even new conventional foods. For example, the kiwi fruit was introduced into the U.S. and the European markets in the 1960s with no known human allergies; however, today there are people allergic to this fruit."

Genetic modification can also be used to remove allergens from foods, potentially reducing the risk of food allergies. A hypo-allergenic strain of soybean was tested in 2003 and shown to lack the major allergen that is found in the beans. A similar approach has been tried in ryegrass, which produces pollen that is a major cause of hay fever: here a fertile GM grass was produced that lacked the main pollen allergen, demonstrating that hypoallergenic grass is also possible.

The development of genetically modified products found to cause allergic reactions has been halted by the companies developing them before they were brought to market. In the early 1990s, Pioneer Hi-Bred attempted to improve the nutrition content of soybeans intended for animal feed by adding a gene from the Brazil nut. Because they knew that people have allergies to nuts, Pioneer ran in vitro and skin prick allergy tests. The tests showed that the transgenic soy was allergenic. Pioneer Hi-Bred therefore discontinued further development. In 2005, a pest-resistant field pea developed by the Australian Commonwealth Scientific and Industrial Research Organisation for use as a pasture crop was shown to cause an allergic reaction in mice. Work on this variety was immediately halted. These cases have been used as evidence that genetic modification can produce unexpected and dangerous changes in foods, and as evidence that safety tests effectively protect the food supply.

During the Starlink corn recalls in 2000, a variety of GM maize containing the *Bacillus thuringiensis* (Bt) protein Cry9C, was found contaminating corn products in U.S. supermarkets and restaurants. It was also found in Japan and South Korea. Starlink corn had only been approved for animal feed as the Cry9C protein lasts longer in the digestive system than other Bt proteins raising concerns about its potential allergenicity. In 2000, Taco Bell-branded taco shells sold in supermarkets were found to contain Starlink, resulting in a recall of those products, and eventually led to the recall of over 300 products. Sales of StarLink seed were discontinued and the registration for the Starlink varieties was voluntarily withdrawn by Aventis in October 2000. Aid sent by the United Nations and the United States to Central African nations was also found to be contaminated with StarLink corn and the aid was rejected. The U.S. corn supply has been monitored for Starlink Bt proteins since 2001 and no positive samples have been found since 2004. In response, GeneWatch UK and Greenpeace set up the GM Contamination Register in 2005. During the recall, the United States Centers for Disease Control evaluated reports of allergic reactions to StarLink corn, and determined that no allergic reactions to the corn had occurred.

Horizontal Gene Transfer

Horizontal gene transfer is the movement of genes from one organism to another in a manner other than reproduction.

The risk of horizontal gene transfer between GMO plants and animals is very low and in most cases is expected to be lower than background rates. Two studies on the possible effects of feeding animals with genetically modified food found no residues of recombinant DNA or novel proteins in any organ or tissue samples. Studies found DNA from the M13 virus, Green fluorescent protein and Rubisco genes in the blood and tissue of animals, and in 2012, a paper suggested that a specific microRNA from rice could be found at very low quantities in human and animal serum. Other studies however, found no or negligible transfer of plant microRNAs into the blood of humans or any of three model organisms.

Another concern is that the antibiotic resistance gene commonly used as a genetic marker in transgenic crops could be transferred to harmful bacteria, creating resistant superbugs. A 2004 study involving human volunteers examined whether the transgene from modified soy would transfer to bacteria that live in the human gut. As of 2012 it was the only human feeding study to have been conducted with GM food. The transgene was detected in three volunteers from a group of seven who had previously had their large intestines removed for medical reasons. As this gene transfer did not increase after the consumption of the modified soy, the researchers concluded that gene transfer did not occur. In volunteers with intact digestive tracts, the transgene did not survive. The antibiotic resistance genes used in genetic engineering are naturally found in many pathogens and antibiotics these genes confer resistance to are not widely prescribed.

Animal Feeding Studies

Reviews of animal feeding studies mostly found no effects. A 2014 review found that the performance of animals fed GM feed was similar to that of animals fed "isogenic non-GE crop lines". A 2012 review of 12 long-term studies and 12 multigenerational studies conducted by public research laboratories concluded that none had discovered any safety problems linked to consumption of GM food. A 2009 review by Magaña-Gómez found that although most studies concluded that modified foods do not differ in nutrition or cause toxic effects in animals, some did report adverse changes at a cellular level caused by specific modified foods. The review concluded that

"More scientific effort and investigation is needed to ensure that consumption of GM foods is not likely to provoke any form of health problem". Dona and Arvanitoyannis' 2009 review concluded that "results of most studies with GM foods indicate that they may cause some common toxic effects such as hepatic, pancreatic, renal, or reproductive effects and may alter the hematological, biochemical, and immunologic parameters". Reactions to this review in 2009 and 2010 noted that Dona and Arvanitoyannis had concentrated on articles with an anti-modification bias that were refuted in peer-reviewed articles elsewhere. Flachowsky concluded in a 2005 review that food with a one-gene modification were similar in nutrition and safety to non-modified foods, but he noted that food with multiple gene modifications would be more difficult to test and would require further animal studies. A 2004 review of animal feeding trials by Aumaitre and others found no differences among animals eating genetically modified plants.

In 2007, Domingo's search of the Pubmed database using 12 search terms indicated that the "number of references" on the safety of GM or transgenic crops was "surprisingly limited," and he questioned whether the safety of GM food had been demonstrated. The review also stated that its conclusions were in agreement with three earlier reviews. However, Vain found 692 research studies in 2007 that focused on GM crop and food safety and found increasing publication rates of such articles in recent years. Vain commented that the multidisciplinarian nature of GM research complicated the retrieval of studies based on it and required many search terms (he used more than 300) and multiple databases. Domingo and Bordonaba reviewed the literature again in 2011 and said that, although there had been a substantial increase in the number of studies since 2006, most were conducted by biotechnology companies "responsible of commercializing these GM plants."

Human Studies

While some groups and individuals have called for more human testing of GM food, multiple obstacles complicate such studies. The General Accounting Office (in a review of FDA procedures requested by Congress) and a working group of the Food and Agricultural and World Health organizations both said that long-term human studies of the effect of GM food are not feasible. The reasons included lack of a plausible hypothesis to test, lack of knowledge about the potential long-term effects of conventional foods, variability in the ways humans react to foods and that epidemiological studies were unlikely to differentiate modified from conventional foods, which come with their own suite of unhealthy characteristics.

Additionally, ethical concerns guide human subject research. These mandate that each tested intervention must have a potential benefit for the human subjects, such as treatment for a disease or nutritional benefit (ruling out, e.g., human toxicity testing). Kimber claimed that the "ethical and technical constraints of conducting human trials, and the necessity of doing so, is a subject that requires considerable attention." Food with nutritional benefits may escape this objection. E.g., GM rice has been tested for nutritional benefits, namely, increased levels of Vitamin A.

Controversial Studies

Pusztai Affair

Arpad Pusztai published the first peer-reviewed paper to find negative effects from GM food consumption in 1999. Pusztai fed rats potatoes transformed with the *Galanthus nivalis* agglutinin

(GNA) gene from the *Galanthus* (snowdrop) plant, allowing the tuber to synthesise the GNA lectin protein. While some companies were considering growing GM crops expressing lectin, GNA was an unlikely candidate. Lectin is toxic, especially to gut epithelia. Pusztai reported significant differences in the thickness of the gut epithelium, but no differences in growth or immune system function.

On June 22, 1998, an interview on Granada Television's current affairs programme World in Action, Pusztai said that rats fed on the potatoes had stunted growth and a repressed immune system. A media frenzy resulted. Pusztai was suspended from the Rowett Institute. Misconduct procedures were used to seize his data and ban him from speaking publicly. The Rowett Institute and the Royal Society reviewed his work and concluded that the data did not support his conclusions. The work was criticized on the grounds that the unmodified potatoes were not a fair control diet and that any rat fed only potatoes would suffer from protein deficiency. Pusztai responded by stating that all diets had the same protein and energy content and that the food intake of all rats was the same.

Bt Corn

A 2011 study was the first to evaluate the correlation between maternal and fetal exposure to Bt toxin produced in GM maize and to determine exposure levels of the pesticides and their metabolites. It reported the presence of pesticides associated with the modified foods in women and in pregnant women's fetuses. The paper and related media reports were criticized for overstating the results. Food Standards Australia New Zealand (FSANZ) posted a direct response, saying that the suitability of the ELISA method for detecting the Cry1Ab protein was not validated and that no evidence showed that GM food was the protein's source. The organization also suggested that even had the protein been detected its source was more likely conventional or organic food.

Séralini Affair

In 2007, 2009, and 2011, Gilles-Éric Séralini published re-analysis studies that used data from Monsanto rat-feeding experiments for three modified maize varieties (insect-resistant MON 863 and MON 810 and glyphosate-resistant NK603). He concluded that the data showed liver, kidney and heart damage. The European Food Safety Authority (EFSA) then concluded that the differences were all within the normal range. EFSA also stated that Séralini's statistics were faulty. EFSA's conclusions were supported by FSANZ, a panel of expert toxicologists, and the French High Council of Biotechnologies Scientific Committee (HCB).

In 2012, Séralini's lab published a paper that considered the long-term effects of feeding rats various levels of GM glyphosate-resistant maize, conventional glyphosate-treated maize, and a mixture of the two strains. The paper concluded that rats fed the modified maize had severe health problems, including liver and kidney damage and large tumors. The study provoked widespread criticism. Séralini held a press conference just before the paper was released in which he announced the release of a book and a movie. He allowed reporters to have access to the paper before his press conference only if they signed a confidentiality agreement under which they could not report other scientists' responses to the paper. The press conference resulted in media coverage emphasizing a connection between GMOs, glyphosate, and cancer. Séralini›s publicity stunt yielded criticism from other scientists for prohibiting critical commentary. Criticisms included insufficient statistical power and that Séralini's Sprague-Dawley rats were inappropriate for a lifetime study (as opposed to a shorter toxicity study) because of their tendency to develop cancer (one study found that more than 80% normally got cancer). The Organisation for Economic Co-operation and Development guidelines recommended using 65 rats per experiment instead of the 10 in Séralini's. Other criticisms included the lack of data regarding food amounts and specimen growth rates, the lack of a dose–response relationship (females fed three times the standard dose showed a decreased number of tumours) and no identified mechanism for the tumour increases. Six French national academies of science issued an unprecedented joint statement condemning the study and the journal that published it. *Food and Chemical Toxicology* published many critical letters, with only a few expressing support. National food safety and regulatory agencies also reviewed the paper and dismissed it. In March 2013, Séralini responded to these criticisms in the same journal that originally published his study, and a few scientists supported his work. In November 2013, the editors of *Food and Chemical Toxicology* retracted the paper. The retraction was met with protests from Séralini and his supporters. In 2014, the study was republished by a different journal, *Environmental Sciences Europe*, in an expanded form, including the raw data that Séralini had originally refused to reveal.

Environment

Genetically modified crops are planted in fields much like regular crops. There they interact directly with organisms that feed on the crops and indirectly with other organisms in the food chain. The pollen from the plants is distributed in the environment like that of any other crop. This distribution has led to concerns over the effects of GM crops on the environment. Potential effects include gene flow, pesticide resistance and greenhouse gas emissions.

Non-Target Organisms

A major use of GM crops is in insect control through the expression of the *cry* (crystal delta-endotoxins) and *Vip* (vegetative insecticidal proteins) genes from *Bacillus thuringiensis* (Bt). Such toxins could affect other insects in addition to targeted pests such as the European corn borer. Bt proteins have been used as organic sprays for insect control in France since 1938 and the US since 1958, with no reported ill effects. *Cry* proteins selectively target Lepidopterans (moths and butterflies). As a toxic mechanism, *cry* proteins bind to specific receptors on the membranes of mid-gut (epithelial) cells, resulting in their rupture. Any organism that lacks the appropriate receptors in its gut is unaffected by the *cry* protein, and therefore is not affected by Bt. Regulatory agencies assess the potential for transgenic plants to affect non-target organisms before approving their commercial release.

In 1999 a paper stated that, in a laboratory environment, pollen from Bt maize dusted onto milkweed could harm the monarch butterfly. A collaborative research exercise over the following two years by several groups of scientists in the US and Canada studied the effects of Bt pollen in both the field and the laboratory. The study resulted in a risk assessment concluding that any risk posed to butterfly populations was negligible. A 2002 review of the scientific literature concluded that "the commercial large-scale cultivation of current Bt–maize hybrids did not pose a significant risk to the monarch population" and noted that despite large-scale planting of genetically modified crops, the butterfly's population was increasing. However, the herbicide glyphosate used to grow

GMOs kills milkweed, the only food source of Monarch butterflies, and by 2015 about 90% of the U.S. population has declined.

Lövei et al. analyzed laboratory settings and found that Bt toxins could affect non-target organisms, generally closely related to the intended targets. Typically, exposure occurs through the consumption of plant parts, such as pollen or plant debris, or through Bt ingestion by predators. A group of academic scientists criticized the analysis, writing: "We are deeply concerned about the inappropriate methods used in their paper, the lack of ecological context, and the authors' advocacy of how laboratory studies on non-target arthropods should be conducted and interpreted".

Biodiversity

Crop genetic diversity might decrease due to the development of superior GM strains that crowd others out of the market. Indirect effects might affect other organisms. To the extent that agrochemicals impact biodiversity, modifications that increase their use, either because successful strains require them or because the accompanying development of resistance will require increased amounts of chemicals to offset increased resistance in target organisms.

Studies comparing the genetic diversity of cotton found that in the US diversity has either increased or stayed the same, while in India it has declined. This difference was attributed to the larger number of modified varieties in the US compared to India. A review of the effects of Bt crops on soil ecosystems found that in general they "appear to have no consistent, significant, and long-term effects on the microbiota and their activities in soil".

The diversity and number of weed populations has been shown to decrease in farm-scale trials in the United Kingdom and in Denmark when comparing herbicide-resistant crops to their conventional counterparts. The UK trial suggested that the diversity of birds could be adversely affected by the decrease in weed seeds available for foraging. Published farm data involved in the trials showed that seed-eating birds were more abundant on conventional maize after the application of the herbicide, but that there were no significant differences in any other crop or prior to herbicide treatment. A 2012 study found a correlation between the reduction of milkweed in farms that grew glyphosate-resistant crops and the decline in adult monarch butterfly populations in Mexico. *The New York Times* reported that the study "raises the somewhat radical notion that perhaps weeds on farms should be protected.

A 2005 study designed to "simulate the impact of a direct overspray on a wetland" with four different agrochemicals (carbaryl (Sevin), malathion, 2,4-dichlorophenoxyacetic acid, and glyphosate in a Roundup formulation) by creating artificial ecosystems in tanks and then applying "each chemical at the manufacturer's maximum recommended application rates" found that "species richness was reduced by 15% with Sevin, 30% with malathion, and 22% with Roundup, whereas 2,4-D had no effect". The study has been used by environmental groups to argue that use of agrochemicals causes unintended harm to the environment and to biodiversity.

Secondary Pests

Several studies documented surges in secondary pests within a few years of adoption of Bt cotton. In China, the main problem has been with mirids, which have in some cases "completely eroded all benefits from Bt cotton cultivation". A 2009 study in China concluded that the increase in secondary pests depended on local temperature and rainfall conditions and occurred in half the villages studied. The increase in insecticide use for the control of these secondary insects was far smaller than the reduction in total insecticide use due to Bt cotton adoption. A 2011 study based on a survey of 1,000 randomly selected farm households in five provinces in China found that the reduction in pesticide use in Bt cotton cultivars was significantly lower than that reported in research elsewhere: The finding was consistent with a hypothesis that more pesticide sprayings are needed over time to control emerging secondary pests, such as aphids, spider mites, and lygus bugs. Similar problems have been reported in India, with mealy bugs and aphids.

Gene Flow

Genes from a GMO may pass to another organism just like an endogenous gene. The process is known as outcrossing and can occur in any new open-pollinated crop variety. Introduced traits potentially can cross into neighboring plants of the same or closely related species through three different types of gene flow: crop-to-crop, crop-to-weedy, and crop-to-wild. In crop-to-crop, genetic information from a genetically modified crop is transferred to a non-genetically modified crop. Crop-to-weedy transfer refers to the transfer of genetically modified material to a weed, and crop-to-wild indicates transfer from a genetically modified crop to a wild, undomesticated plant and/or crop. There are concerns that the spread of genes from modified organisms to unmodified relatives could produce species of weeds resistant to herbicides that could contaminate nearby non-genetically modified crops, or could disrupt the ecosystem, This is primarily a concern if the transgenic organism has a significant survival capacity and can increase in frequency and persist in natural populations. This process, whereby genes are transferred from GMOs to wild relatives, is different from the development of so-called "superweeds" or "superbugs" that develop resistance to pesticides under natural selection.

In most countries environmental studies are required before approval of a GMO for commercial purposes, and a monitoring plan must be presented to identify unanticipated gene flow effects.

In 2004, Chilcutt and Tabashnik found Bt protein in kernels of in a refuge (a conventional crop planted to harbor pests that might otherwise become resistant a pesticide associated with the GMO) implying that gene flow had occurred.

In 2005, scientists at the UK Centre for Ecology and Hydrology reported the first evidence of horizontal gene transfer of pesticide resistance to weeds, in a few plants from a single season; they found no evidence that any of the hybrids had survived in subsequent seasons.

In 2007, the U.S. Department of Agriculture fined Scotts Miracle-Gro $500,000 when modified DNA from GM creeping bentgrass, was found within relatives of the same genus (*Agrostis*) as well as in native grasses up to 21 km (13 mi) from the test sites, released when freshly cut, wind-blown grass.

In 2009 Mexico created a regulatory pathway for GM maize, but because Mexico is maize's center of diversity, concerns were raised about GM maize's effects on local strains. A 2001 report found

Bt maize cross-breeding with conventional maize in Mexico. The data in this paper was later described as originating from an artifact and the publishing journal *Nature* stated that "the evidence available is not sufficient to justify the publication of the original paper", although it did not retract the paper. A subsequent large-scale study, in 2005, found no evidence of gene flow in Oaxaca. However, other authors claimed to have found evidence of such gene flow.

A 2010 study showed that about 83 percent of wild or weedy canola tested contained genetically modified herbicide resistance genes. According to the researchers, the lack of reports in the United States suggested that oversight and monitoring were inadequate. A 2010 report stated that the advent of glyphosate-resistant weeds could cause GM crops to lose their effectiveness unless farmers combined glyphosate with other weed-management strategies.

One way to avoid environmental contamination is genetic use restriction technology (GURT), also called "Terminator". This uncommercialized technology would allow the production of crops with sterile seeds, which would prevent the escape of GM traits. Groups concerned about food supplies had expressed concern that the technology would be used to limit access to fertile seeds. Another hypothetical technology known as "Traitor" or "T-GURT", would not render seeds sterile, but instead would require application of a chemical to GM crops to activate engineered traits. Groups such as Rural Advancement Foundation International raised concerns that further food safety and environmental testing needed to be done before T-GURT would be commercialized.

Escape of Modified Crops

The escape of genetically modified seed into neighboring fields, and the mixing of harvested products, is of concern to farmers who sell to countries that do not allow GMO imports.

In 1999 scientists in Thailand claimed they had discovered unapproved glyphosate-resistant GM wheat in a grain shipment, even though it was only grown in test plots. No mechanism for the escape was identified.

In 2000, Aventis StarLink GM corn was found in US markets and restaurants. It became the subject of a recall that started when Taco Bell-branded taco shells sold in supermarkets were found to contain it. StarLink was then discontinued. Registration for Starlink varieties was voluntarily withdrawn by Aventis in October 2000.

American rice exports to Europe were interrupted in 2006 when the LibertyLink modification was found in commercial rice crops, although it had not been approved for release. An investigation by the USDA's Animal and Plant Health Inspection Service (APHIS) failed to determine the cause of the contamination.

In May 2013, unapproved glyphosate-resistant GM wheat (but that had been approved for human consumption) was discovered in a farm in Oregon in a field that had been planted with winter wheat. The strain was developed by Monsanto, and had been field-tested from 1998 to 2005. The discovery threatened US wheat exports which totaled $8.1 billion in 2012. Japan, South Korea and Taiwan temporarily suspended winter wheat purchases as a result of the discovery. As of August 30, 2013, while the source of the modified wheat remained unknown, Japan, South Korea and Taiwan had resumed placing orders.

Coexistence with Conventional Crops

The US has no legislation governing the relationship among mixtures of farms that grow organic, conventional and GM crops. The country relies on a "complex but relaxed" combination of three federal agencies (FDA, EPA, and USDA/APHIS) and states' common law tort systems to manage coexistence. The Secretary of Agriculture convened an Advisory Committee on Biotechnology and 21st Century Agriculture (AC21) to study coexistence and make recommendations about the issue. The members of AC21 included representatives of the biotechnology industry, the organic food industry, farming communities, the seed industry, food manufacturers, State governments, consumer and community development groups, the medical profession, and academic researchers. AC21 recommended that a study assess the potential for economic losses to US organic farmers; that any serious losses lead to a crop insurance program, an education program to ensure that organic farmers put appropriate contracts in place and that neighboring GMO farmers take appropriate containment measures. Overall the report supported a diverse agriculture system supporting diverse farming systems.

The EU implemented regulations specifically governing co-existence and traceability. Traceability has become commonplace in the food and feed supply chains of most countries, but GMO traceability is more challenging given strict legal thresholds for unwanted mixing. Since 2001, conventional and organic food and feedstuffs can contain up to 0.9% of authorised modified material without carrying a GMO label. (any trace of non-authorised modification is cause for a shipment to be rejected). Authorities require the ability to trace, detect and identify GMOs, and the several countries and interested parties created a non-governmental organization, Co-Extra, to develop such methods.

Chemical Use

Birth defects in Hawaii caused a controversy, and public outcry of people concerned over heavy pesticide usage. Hawaii uses ten times the amount of pesticides compared to the rest of the U.S.

Pesticides

Pesticides destroy, repel or mitigate pests (an organism that attacks or competes with a crop). A 2014 meta-analysis covering 147 original studies of farm surveys and field trials, and 15 studies from the researchers conducting the study, concluded that adoption of GM technology had reduced chemical pesticide use by 37%, with the effect larger for insect-tolerant crops than herbicide-tolerant crops.

Herbicides

The development of glyphosate-tolerant (Roundup Ready) plants changed the herbicide use profile away from more persistent, higher toxicity herbicides, such as atrazine, metribuzin and alachlor, and reduced the volume and harm of herbicide runoff. A study by Chuck Benbrook concluded that the spread of glyphosate-resistant weeds had increased US herbicide use. That study cited a 23% increase (.3 kilograms/hectare) for soybeans from 1996-2006, a 43% (.9 kg/ha) increase for cotton from 1996-2010 and a 16% (.5 kg/ha) decrease for corn from 1996-2010. However, this study came under scrutiny because Benbrook did not consider the fact that glyphosate is less toxic than other herbicides, thus net toxicity may decrease even as use increases. Graham Brookes accused Benbrook of subjective herbicide estimates because his data, provided by the National Agricultural Statistics Service, does not distinguish between genetically modified and non-genetically modified crops. Brookes had earlier published a study that found that the use of biotech crops had reduced the volume and environmental impact of herbicide and other pesticides, which contradicted Benbrook. Brookes stated that Benbrook had made "biased and inaccurate" assumptions.

Insecticides

A claimed environmental benefit of Bt-cotton and maize is reduced insecticide use. A PG Economics study concluded that global pesticide use was reduced by 286,000 tons in 2006, decreasing pesticidal environmental impact by 15%. A survey of small Indian farms between 2002 and 2008 concluded that Bt cotton adoption had led to higher yields and lower pesticide use. Another study concluded that insecticide use on cotton and corn during the years 1996 to 2005 fell by 35,600,000 kilograms (78,500,000 lb) of active ingredient, roughly equal to the annual amount applied in the European Union. A Bt cotton study in six northern Chinese provinces from 1990 to 2010 concluded that it halved the use of pesticides and doubled the level of ladybirds, lacewings and spiders and extended environmental benefits to neighbouring crops of maize, peanuts and soybeans.

Resistant Insect Pests

Resistance evolves naturally after a population has been subjected to selection pressure via repeated use of a single pesticide. In November 2009, Monsanto scientists found that the pink bollworm had become resistant to first generation Bt cotton in parts of Gujarat, India—that generation expresses one Bt gene, *Cry1Ac*. This was the first instance of Bt resistance confirmed by Monsanto. Similar resistance was later identified in Australia, China, Spain and the US.

One strategy to delay Bt-resistance is to plant pest refuges using conventional crops, thereby diluting any resistant genes. Another is to develop crops with multiple Bt genes that target different receptors within the insect. In 2012, a Florida field trial demonstrated that army worms were resistant to Dupont-Dow's GM corn. This resistance was discovered in Puerto Rico in 2006, prompting Dow and DuPont to stop selling the product there. The European corn borer, one of Bt's primary targets, is also capable of developing resistance.

Economy

GM food's economic value to farmers is one of its major benefits, including in developing nations. A 2010 study found that Bt corn provided economic benefits of $6.9 billion over the previous 14 years in five Midwestern states. The majority ($4.3 billion) accrued to farmers producing non-Bt corn. This was attributed to European corn borer populations reduced by exposure to Bt corn, leaving fewer to attack conventional corn nearby. Agriculture economists calculated that "world surplus [increased by] $240.3 million for 1996. Of this total, the largest share (59%) went to U.S. farmers. Seed company Monsanto received the next largest share (21%), followed by US consumers (9%), the rest of the world (6%), and the germplasm supplier, Delta and Pine Land Company (5%)." PG Economics comprehensive 2012 study concluded that GM crops increased farm incomes worldwide by $14 billion in 2010, with over half this total going to farmers in developing countries.

The main Bt crop grown by small farmers in developing countries is cotton. A 2006 review of Bt cotton findings by agricultural economists concluded, “the overall balance sheet, though promising, is mixed. Economic returns are highly variable over years, farm type, and geographical location”. However, environmental activist Mark Lynas said that complete rejection of genetic engineering is “illogical and potentially harmful to the interests of poorer peoples and the environment”.

In 2013 the European Academies Science Advisory Council (EASAC) asked the EU to allow the development of agricultural GM technologies to enable more sustainable agriculture, by employing fewer land, water and nutrient resources. EASAC also criticizes the EU’s “timeconsuming and expensive regulatory framework” and said that the EU had fallen behind in the adoption of GM technologies.

Industrial Agriculture

GM crops play a key role in intensive crop farming, which involves monoculture, use of herbicides and pesticides, use of equipment requiring large amounts of fuel and irrigation. Opponents such as Jonathan Latham of the *Bioscience Resource Project* and Vandana Shiva treat industrial agriculture and modified crops as closely related topics, and call for agriculture to adopt practices drastically reducing disruptions to the environment.

Proponents of conventional agriculture point to its high yields, low prices, and wider choices and claim that technology is necessary to feed a growing world population.

Developing Nations

Disagreements about developing nations include the claimed need for increased food supplies, and how to achieve such an increase. Some scientists suggest that a second Green Revolution including use of modified crops is needed to provide sufficient food. The potential for genetically modified food to help developing nations was recognised by the International Assessment of Agricultural Science and Technology for Development, but as of 2008 they had found no conclusive evidence of a solution.

Skeptics such as John Avise claim that apparent shortages are caused by problems in food distribution and politics, rather than production. Other critics say that the world has so many people because the second green revolution adopted unsustainable agricultural practices that left the world with more mouths to feed than the planet can sustain. Pfeiffer claimed that even if technological farming could feed the current population, its dependence on fossil fuels, which in 2006 he incorrectly predicted would reach peak output in 2010, would lead to a catastrophic rise in energy and food prices.

Claimed deployment constraints to developing nations include the lack of easy access, equipment costs and intellectual property rights that hurt developing countries. The Consultative Group on International Agricultural Research (CGIAR), an aid and research organization, was praised by the World Bank for its efforts, but the bank recommended that they shift to genetics research and productivity enhancement. Obstacles include access to patents, commercial licenses and the difficulty that developing countries have in accessing genetic resources and other intellectual property. The International Treaty on Plant Genetic Resources for Food and Agriculture attempted to remedy this problem, but results have been inconsistent. As a result, “orphan crops”, such as teff, millets, cowpeas and indigenous plants, which are important in these countries receive little investment.

Writing about Norman Borlaug’s 2000 publication *Ending world hunger: the promise of biotechnology and the threat of antiscience zealotry*, the authors argued that Borlaug’s warnings were still true in 2010:

GM crops are as natural and safe as today’s bread wheat, opined Dr. Borlaug, who also reminded agricultural scientists of their moral obligation to stand up to the antiscience crowd and warn policy makers that global food insecurity will not disappear without this new technology and ignoring this reality would make future solutions all the more difficult to achieve.

Yield

US maize yields were flat until the 1930s, when the adoption of conventional hybrid seeds caused them to increase by ~.8 bushels/acre (1937-1955). Thereafter a combination of improved genetics, fertilizer and pesticide availability and mechanization raised the rate of increase to 1.9 bushels per acre per year. In the years since the advent of GM maize, the rate increased slightly to 2.0. Average US maize yields were 174.2 bushels per acre in 2014.

Commercial GM crops have traits that reduce yield loss from insect pressure or weed interference.

2014 Review

A 2014 review concluded that GM crops’ effects on farming were positive. According to *The Economist*, the meta-analysis considered all published English-language examinations of the agronomic and economic impacts between 1995 and March 2014. The study found that herbicide-tolerant crops have lower production costs, while for insect-resistant crops the reduced pesticide use was offset by higher seed prices, leaving overall production costs about the same.

Yields increased 9% for herbicide tolerance and 25% for insect resistance. Farmers who adopted GM crops made 69% higher profits than those who did not. The review found that GM crops help farmers in developing countries, increasing yields by 14 percentage points.

The researchers considered some studies that were not peer-reviewed, and a few that did not report sample sizes. They attempted to correct for publication bias, by considering sources beyond academic journals. The large data set allowed the study to control for potentially confounding variables such as fertiliser use. Separately, they concluded that the funding source did not influence study results.

2010 Review

A 2010 article supported by CropLife International summarised the results of 49 peer reviewed studies. On average, farmers in developed countries increased yields by 6% and 29% in developing countries.

Tillage decreased by 25–58% on herbicide-resistant soybeans. Glyphosate-resistant crops allowed farmers to plant rows closer together as they did not have to control post-emergent weeds with mechanical tillage. Insecticide applications on Bt crops were reduced by 14–76%. 72% of farmers worldwide experienced positive economic results.

2009 Review

In 2009 the Union of Concerned Scientists, a group opposed to genetic engineering and cloning of food animals, summarized peer-reviewed studies on the yield contribution of GM soybeans and maize in the US. The report concluded that other agricultural methods had made a greater contribution to national crop yield increases in recent years than genetic engineering.

Wisconsin Study

A study unusually published as correspondence rather than as an article examined maize modified to express four traits (resistance to European corn borer, resistance to corn root worm, glyphosate tolerance and glyfosinate tolerance) singly and in combination in Wisconsin fields from 1990-2010. The variance in yield from year to year was reduced, equivalent to a yield increase of 0.8-4.2 bushels per acre. Bushel per acre yield changes were +6.4 for European corn borer resistance, +5.76 for glufosinate tolerance, -5.98 for glyphosate tolerance and -12.22 for corn rootworm resistance. The study found interactions among the genes in multi-trait hybrid strains, such that the net effect varied from the sum of the individual effects. For example, the combination of European corn borer resistance and glufosinate tolerance increased yields by 3.13, smaller than either of the individual traits

Market Dynamics

The seed industry is dominated by a small number of vertically integrated firms. In 2011, 73% of the global market was controlled by 10 companies.

In 2001, the USDA reported that industry consolidation led to economies of scale, but noted that the move by some companies to divest their seed operations questioned the long-term viability of these conglomerates. Two economists have said that the seed companies' market power could raise welfare despite their pricing strategies, because "even though price discrimination is often considered to be an unwanted market distortion, it may increase total welfare by increasing total output and by making goods available to markets where they would not appear otherwise."

Market share gives firms the ability to set or influence price, dictate terms, and act as a barrier to entry. It also gives firms bargaining power over governments in policy making. In March 2010, the US Department of Justice and the US Department of Agriculture held a meeting in Ankeny, Iowa, to look at the competitive dynamics in the seed industry. Christine Varney, who heads the antitrust division in the Justice Department, said that her team was investigating whether biotech-seed patents were being abused. A key issue was how Monsanto licenses its patented glyphosate-tolerance trait that was in 93 percent of US soybeans grown in 2009. About 250 family farmers, consumers and other critics of corporate agriculture held a town meeting prior to the government meeting to protest Monsanto's purchase of independent seed companies, patenting seeds and then raising seed prices.

Intellectual Property

Traditionally, farmers in all nations saved their own seed from year to year. However, since the early 1900s hybrid crops have been widely used in the developed world and seeds to grow these crops are purchased each year from seed producers. The offspring of the hybrid corn, while still viable, lose hybrid vigor (the beneficial traits of the parents). This benefit of first-generation hybrid seeds is the primary reason for not planting second-generation seed. However, for non-hybrid GM crops, such as GM soybeans, seed companies use intellectual property law and tangible property common law, each expressed in contracts, to prevent farmers from planting saved seed. For example, Monsanto's typical bailment license (covering transfer of the seeds themselves) forbids saving seeds, and also requires purchasers to sign a separate patent license agreement.

Corporations say that they need to prevent seed piracy, to fulfill financial obligations to shareholders, and to finance further development. DuPont spent approximately half its $2 billion research and development (R&D) budget on agriculture in 2011 while Monsanto spends 9-10% of sales on R&D.

Detractors such as Greenpeace say that patent rights give corporations excessive control over agriculture. The Center for Ecoliteracy claimed that "patenting seeds gives companies excessive power over something that is vital for everyone." A 2000 report stated, "If the rights to these tools are strongly and universally enforced - and not extensively licensed or provided pro bono in the developing world - then the potential applications of GM technologies described previously are unlikely to benefit the less developed nations of the world for a long time" (i.e. until after the restrictions expire).

Monsanto has patented its seed and it obligates farmers who choose to buy its seeds to sign a license agreement, obligating them store or sell, but not plant, all the crops that they grow.

Lawsuits Filed Against Farmers for Patent Infringement

Monsanto has filed patent infringement suits against 145 farmers, but proceeded to trial with only 11. In some of the latter, the defendants claimed unintentional contamination by gene flow, but Monsanto won every case. Monsanto Canada's Director of Public Affairs stated, "It is not, nor has it ever been Monsanto Canada's policy to enforce its patent on Roundup Ready crops when they are present on a farmer's field by accident...Only when there has been a knowing and deliberate violation of its patent rights will Monsanto act." In 2009 Monsanto announced that after its soybean patent expires in 2014, it will no longer prohibit farmers from planting soybean seeds that they grow.

One example of such litigation is the Monsanto v. Schmeiser case. This case is widely misunderstood. In 1997, Percy Schmeiser, a canola breeder and grower in Bruno, Saskatchewan, discovered that one of his fields had canola that was resistant to Roundup. He had not purchased this seed, which had blown onto his land from neighboring fields. He later harvested the area and saved the crop in the back of a pickup truck.Before the 1998 planting, Monsanto representatives informed Schmeiser that using this crop for seed would infringe the patent, and offered him a license, which Schmeiser refused.According to the Canadian Supreme Court, after this conversation "Schmeiser nevertheless took the harvest he had saved in the pick-up truck to a seed treatment plant and had it treated for use as seed. Once treated, it could be put to no other use. Mr. Schmeiser planted the treated seed in nine fields, covering approximately 1,000 acres in all....A series of independent tests by different experts confirmed that the canola Mr. Schmeiser planted and grew in 1998 was

95 to 98 percent Roundup resistant."After further negotiations between Schmeiser and Monsanto broke down, Monsanto sued Schmeiser for patent infringement and prevailed in the initial case. Schmeiser appealed and lost, and appealed again to the Canadian Supreme Court, which in 2004 ruled 5 to 4 in Monsanto's favor, stating that "it is clear on the findings of the trial judge that the appellants saved, planted, harvested and sold the crop from plants containing the gene and plant cell patented by Monsanto."

International Trade

GM crops have been the source of international trade disputes and tensions within food-exporting nations over whether introduction of genetically modified crops would endanger exports to other countries.

In Canada in 2010, flax exports to Europe were rejected when traces of an experimental GM flax were found in shipments. This led a member of Parliament to propose Private Member's Bill C-474, which would have required that "an analysis of potential harm to export markets be conducted before the sale of any new genetically engineered seed is permitted." Opponents claimed that "incorporating stringent socio-economic standards into the science-based regulatory system could spell the end of private research funding; because if private biotechnology companies can't see the possibility of a return on their investment, they'll invest their research budget elsewhere." The bill was defeated in 2011, in a "vote of 176 to 97."

Regulation

Labeling

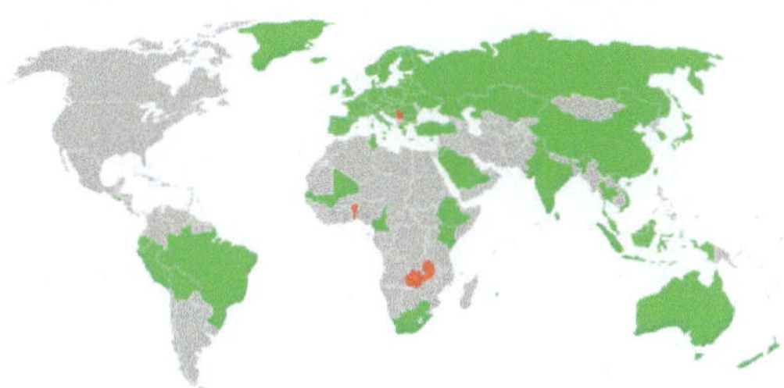

Green: Mandatory labeling required; Red: Ban on import and cultivation of genetically engineered food. Regulations are pending in Israel and the United States.

In 2014, 64 countries required labeling of all GM foods. These include the European Union, Japan, Australia, New Zealand, Russia, China and India. As of March 2015, Israel was in the process of issuing regulations for labeling of food with ingredients from GMOs.

Alaska required labeling of GMO fish and shellfish in 2005, even though no GM fish had been approved by the FDA at the time. A 2014 Vermont law went into effect on July 1, 2016, and some food manufacturers (including General Mills, Mars, Kellogg's, the Campbell Soup Company, PepsiCo, ConAgra, Frito-Lay, and Bimbo Bakeries USA) began distributing products either locally or nation-wide with labels such as "Partially produced with Genetic Engineering". Other manufacturers removed about 3,000 non-compliant products from sale in Vermont. The federal government of the United States passed a law at the end of that month pre-empting all state laws, including Vermont's. The law requires labeling regulations to be issued by July 2018, and allows indirect disclosure such as with a phone number, bar code, or web site. It is unclear whether the rules will require labeling of oils and sugars from GM crops, where the final product does not contain any "genetic material" as mentioned in the law.

Prior to the new federal rules taking effect, while it does require pre-market approval, the U.S. Food and Drug Administration has not required GMO labeling as long as there are no differences in health, environmental safety, and consumer expectations based on the packaging. The federal rules come after GMO labeling was debated in many state legislatures and defeated in popular referendums in Oregon (2002 and 2014), Colorado (2014), California Proposition 37 (2012), and Washington Initiative 522 (2012). Connecticut and Maine had passed laws in 2013 and 2014 respectively, which would have required GMO food labels if Northeast states with a population of at least 20 million had passed similar laws (and for Connecticut, representing at least four states).

Other jurisdictions make such labeling voluntary or have had plans to require labeling. Major GM food crop exporters like the United States (until 2018), Argentina, and Canada have adopted voluntary labeling approaches; China and Brazil have major GM (largely non-food) crops and have adopted mandatory labelling.

Arguments

The American Public Health Association, the British Medical Association and the Public Health Association of Australia support mandatory labeling. The European Commission argued that mandatory labeling and traceability are needed to allow for informed choice, avoid potential misleading of consumers and facilitate the withdrawal of products if adverse effects on health or the environment are discovered. A 2007 study on the effect of labeling laws found that once labeling went into effect, few products continued to contain GM ingredients. The study also found that costs were higher in food-exporting than in food-importing countries.

The American Medical Association (AMA) and the American Association for the Advancement of Science have opposed mandatory labeling absent scientific evidence of harm. The AMA said that even voluntary labeling is misleading unless accompanied by focused consumer education. The AAAS stated that mandatory labeling "can only serve to mislead and falsely alarm consumers".

[Labeling] efforts are not driven by evidence that GM foods are actually dangerous. Indeed, the science is quite clear: crop improvement by the modern molecular techniques of biotechnology is safe. Rather, these initiatives are driven by a variety of factors, ranging from the persistent perception that such foods are somehow "unnatural" and potentially dangerous to the desire to gain competitive advantage by legislating attachment of a label meant to alarm. Another misconception used as a rationale for labeling is that GM crops are untested.

Objectivity of Regulatory Bodies

Groups such as the Union of Concerned Scientists and Center for Food Safety that have expressed concerns about the FDA's lack of a requirement for additional testing for GMO's, lack of required labeling and the presumption that GMO's are "Generally Recognized as Safe" (GRAS), have questioned whether the FDA is too close to companies that seek approval for their products.

Critics in the U.S. protested the appointment of lobbyists to senior positions in the Food and Drug Administration. Michael R. Taylor, a former Monsanto lobbyist, was appointed as a senior adviser to the FDA on food safety in 1991. After leaving the FDA, Taylor became a vice-president of Monsanto. On 7 July 2009, Taylor returned to government as a senior adviser to the FDA Commissioner.

In 2001, when the Starlink corn recall became public, the U.S. Environmental Protection Agency was criticized for being slow to react by Joseph Mendelson III of the Center for Food Safety. He also criticized the EPA and Aventis CropScience for statements at the time of the recall, that indicated they did not anticipate that such a thing would happen.

The Canadian Biotechnology Advisory Committee that reviewed Canada's regulations in 2003 was accused by environmental and citizen groups of not representing the full spectrum of public interests and for being too closely aligned to industry groups.

Most of the Chinese National Biosafety Committee are involved in biotechnology, a situation that led to criticisms that they do not represent a wide enough range of public concerns.

Litigation and Regulation Disputes

United States

Four federal district court suits have been brought against Animal and Plant Health Inspection Service (APHIS), the agency within USDA that regulates genetically modified plants. Two involved field trials (herbicide-tolerant turfgrass in Oregon; pharmaceutical-producing corn and sugar in Hawaii) and two the deregulation of GM alfalfa. and GM sugar beet. APHIS lost all four cases at trial, with the judges ruling they failed to diligently follow the guidelines set out in the National Environmental Policy Act. However, the Supreme Court overturned the nationwide ban on GM alfalfa and an appeal court allowed the partial deregulation of GM sugar beets. After APHIS prepared Environmental Impact Statements for both alfalfa and sugar beets they were approved.

In 2014, Maui County, Hawaii approved an initiative calling for a moratorium on GMO production and research. The initiative specified penalties including fines and jail for knowing violations and did not limit its scope to commercial agriculture. The initiative passed by about 50.2 to 47.9 percent.

On December 15, 2015, the New York Times ran an op-ed titled "Are You Eating Frankenfish?", saying that the United States congress will debate whether genetically engineered salmon should be labeled.

European Union

Until the 1990s, Europe's regulation was less strict than in the U.S. In 1998, the use of MON810, a Bt expressing maize conferring resistance to the European corn borer, was approved for commercial cultivation in Europe. However, in the 1990s a series of unrelated food crises created consumer apprehension about food safety in general and eroded public trust in government oversight. A bovine spongiform encephalopathy outbreak was the most publicized. In 1998, a *de facto* moratorium led to the suspension of approvals of new GMOs in the EU pending the adoption of revised rules.

In the mid-1990s, government approval of some GMO crops in the United States precipitated public concern in Europe and led to a dramatic decrease in American exports to Europe. "Prior to 1997, corn exports to Europe represented about 4% of total US corn exports, generating about $300 million in sales....For example, before 1997, the U.S. sold about 1.75 million tons of corn annually to Spain and Portugal....But in the 1998–99 crop year, Spain bought less than a tenth of the previous year's amount and Portugal bought none at all."

In May 2003, the US and twelve other countries filed a formal complaint with the World Trade Organization that the EU was violating international trade agreements, by blocking imports of US farm products through its ban on GM food. The countries argued that the EU's regulatory process was far too slow and its standards were unreasonable given the scientific evidence showing that the crops were safe. The case was lobbied by Monsanto and France's Aventis, as well as by US agricultural groups such as the National Corn Growers Association. In response, in June 2003, the European Parliament ratified a U.N. biosafety protocol regulating international trade in GM food, and in July agreed to new regulations requiring labeling and traceability, as well as an opt-out provision for individual countries. The approval of new GMOs resumed in May 2004. While GMOs have been approved since then, approvals remain controversial and various countries have utilized opt-out provisions. In 2006, the World Trade Organization ruled that the pre-2004 restrictions had been violations, although the ruling had little immediate effect since the moratorium had already been lifted.

In late 2007, the US ambassador to France recommended "moving to retaliation" to cause "some pain" against France and the European Union in an attempt to fight the French ban and changes in European policy toward genetically modified crops, according to a US government diplomatic cable obtained by WikiLeaks.

- 20 out of 28 European Countries (including Switzerland) said No to GMOs until October 2015.
- These are the countries which have/implement a ban on GMO:
- Austria
- Belgium for the Wallonia region
- Britain for Scotland, Wales and Northern Ireland
- Bulgaria
- Croatia
- Cyprus
- Denmark
- France
- Germany (except for research)
- Greece
- Hungary

- Italy
- Latvia
- Lithuania
- Luxembourg
- Malta
- Netherlands
- Philippines
- Poland
- Slovenia
- Switzerland

Australia

In May 2014, the Supreme Court of the Australian state of Western Australia dismissed "Marsh v. Baxter". The plaintiff was Steve Marsh, an organic farmer, and the defendant was Michael Baxter, his lifelong neighbor, who grew GM canola. In late 2010 Marsh found seeds from Baxter's crop in his fields. Later, Marsh found escaped GM canola growing amidst his crop. Marsh reported the seed and plants to his local organic certification board, and lost the organic certification of some 70 per cent of his 478 hectare farm. Marsh sued on the grounds that Baxter used a method of harvesting his crop that was substandard and negligent, and on the basis that his land had been widely contaminated. In its summary judgment, the court found that approximately 245 cut canola plants were blown by the wind into Marsh's property, Eagle's Rest. However, Baxter's method (swathing) was "orthodox and well accepted harvest methodology". "In 2011, eight GM canola plants were found to have grown up as self-sown volunteer plants on Eagle Rest," which "were identified and pulled out", and "no more volunteer RR canola plants grew on Eagle Rest in subsequent years." The summary judgment stated that the loss of organic certification "was occasioned by the erroneous application of governing NASAA Standards applicable to NASAA organic operators as regards GMOs (genetically modified organisms) at the time." and that "[t]he absence of a reliable underlying evidentiary platform to support a perpetual injunction against swathing was a significant deficiency".

On June 18, 2014 Marsh announced that he had filed an appeal. One ground was the costs of $803,989 awarded against him. The appeal hearing commenced on 23 March 2015 and was adjourned on 25 March "to deal with an order to ascertain whether Mr Baxter's defence has been financially supported by GM-seed supplier Monsanto and/or the Pastoralists and Graziers Association (PGA)".

Philippines

A petition filed May 17, 2013 by environmental group Greenpeace Southeast Asia and farmer-scientist coalition Masipag (Magsasaka at Siyentipiko sa Pagpapaunlad ng Agrikultura) asked the appellate court to stop the planting of Bt eggplant in test fields, saying the impacts of such an undertaking to the environment, native crops and human health are still unknown. The Court of Appeals granted the petition, citing the precautionary principle stating "when human activities may lead to threats of serious and irreversible damage to the environment that is scientifically plausible but uncertain, actions shall be taken to avoid or diminish the threat." Respondents filed a motion for reconsideration in June 2013 and on September 20, 2013 the Court of Appeals chose to uphold their May decision saying the bt talong field trials violate the people's constitutional right to a "balanced and healthful ecology." The Supreme Court on Tuesday, December 8, 2015 permanently stopped the field testing for Bt (*Bacillus thuringiensis*) talong (eggplant), upholding the decision of the Court of Appeals which stopped the field trials for the genetically modified eggplant.

Innovation in Technology and Regulatory law

The first genetically modified crops were made with transgenic approaches, introducing foreign genes and sometimes using bacteria to transfer the genes. In the US, these foreign genetic elements placed the resulting plant under the jurisdiction of the USDA under the Plant Protection Act. However, as of 2010, newer genetic engineering technologies like genome editing have allowed scientists to modify plant genomes without adding foreign genes, thus escaping USDA regulation. Critics have called for regulation to be changed to keep up with changing technology.

Legislation

African Controversies

In 2002, in the midst of a famine, Zambia refused emergency food aid that contained food from genetically modified crops, based on the precautionary principle.

During a conference in the Ethiopian capital of Addis Ababa, Kingsley Amoako, Executive Secretary of the United Nations Economic Commission for Africa (UNECA), encouraged African nations to accept GM food and expressed dissatisfaction in the public's negative opinion of biotechnology.

Studies for Uganda showed that transgenic bananas had a high potential to reduce rural poverty but that urban consumers with a relatively higher income might reject them.

Critics claimed that shipment of US food to southern Africa was more about promoting the adoption of biotech crops in the region than about hunger. The US was supplying Africa with meals and support during a food crisis they were facing in the early 2000s. However, once some of the African countries realized that these shipments contained GM maize, they rejected the shipments and stopped releasing the food that had been sent to them. Critics accused the US of "exploiting the Southern African famine as a public relations tool". The U.S. countered these comments by saying that European nations were letting millions of Africans suffer from hunger and starvation because of "irrational fears over hypothetical and unproven risks". The US had a pre-GMO policy of shipping US crops as food aid, rather than buying crops in/near the countries that needed aid. The US policy was claimed to be more costly than Europe's.

Genetically modified food controversies in Ghana have been widespread since 2013.

Indian Controversies

India is an agrarian country with around 60% of its people depending directly or indirectly upon agriculture. From 1995 to 2013, a total of 296,438 farmers have killed themselves in India, or an average of 16,469 suicides per year. During the same period, about 9.5 million people died per year in India from other causes including malnutrition, diseases and suicides that were non-farming related, or about 171 million deaths from 1995 to 2013. Activists and scholars have offered a number of conflicting reasons for farmer suicides, such as monsoon failure, high debt burdens, genetically modified crops, government policies, public mental health, personal issues and family problems. There are also accusations of states reporting inaccurate data on farmer suicides.

In India, GM cotton yields in Maharashtra, Karnataka, and Tamil Nadu resulted in an average 42% increase in yield in 2002, the first year of commercial planting. A severe drought in Andhra Pradesh that year prevented any increase in yield, because the GM strain was not drought tolerant. Drought-tolerant variants were later developed. Driven by substantially reduced losses to insect predation, by 2011 88% of Indian cotton was modified. Though disputed, the economic and environmental benefits of GM cotton to farmers have been documented. A study from 2002 through 2008 on the economic impacts of Bt cotton in India, showed that Bt cotton increased yields, profits and living standards of smallholder farmers. However, recently cotton bollworm has been developing resistance to Bt cotton. Consequently, in 2012 Maharashtra banned Bt cotton and ordered an independent socioeconomic study of its use. Indian regulators cleared the Bt brinjal, a genetically modified eggplant, for commercialisation in October 2009. After opposition by some scientists, farmers and environmental groups, a moratorium was imposed on its release in February 2010 "for as long as it is needed to establish public trust and confidence".

As of 1 January 2013, all foods containing GMOs must be labelled. The Legal Metrology (Packaged Commodities) Rules, 2011 states that "every package containing the genetically modified food shall bear at the top of its principal display panel the letters 'GM.'" The rules apply to 19 products including biscuits, breads, cereals and pulses, and a few others. The law faced criticism from consumer rights activists as well as from the packaged-food industry; both sides had major concerns that no logistical framework or regulations had been established to guide the law's implementation and enforcement. On March 21, 2014, the Indian government revalidated 10 GM-based food crops and allowed field trials of GM food crops, including wheat, rice and maize.

References

- Séralini G (2012). Tous Cobayes !: OGM, pesticides et produits chimiques [All Guinea Pigs: GMOs, pesticides and chemicals] (in French). Editions Flammarion. ISBN 9782081262362.
- Pollack M, Shaffer G (2009). When Cooperation Fails: the international law and politics of genetically modified foods. Oxford University Press. ISBN 978-0-19-956705-8.
- Final Report of the PABE research project (December 2001). "Public Perceptions of Agricultural Biotechnologies in Europe". Commission of European Communities. Retrieved February 24, 2016.
- Bashshur, Ramona (February 2013). "FDA and Regulation of GMOs". American Bar Association. Retrieved February 24, 2016.
- Cowan, Tadlock (18 Jun 2011). "Agricultural Biotechnology: Background and Recent Issues" (PDF). Congressional Research Service (Library of Congress). pp. 33–38. Retrieved 27 September 2015.
- "Questions & Answers on Food from Genetically Engineered Plants". US Food and Drug Administration. 22 Jun 2015. Retrieved 29 September 2015.
- Webster, TM; Grey, TL (2015). "Glyphosate-Resistant Palmer Amaranth (Amaranthus palmeri) Morphology, Growth, and Seed Production in Georgia.". Weed Science. 63 (1): 264–272. doi:10.1614/ws-d-14-00051.1.

- Commissioner, Office of the. "Press Announcements - FDA takes several actions involving genetically engineered plants and animals for food". www.fda.gov. Retrieved 2015-12-03.
- Chokshi, Niraj (9 May 2014). "Vermont just passed the nation's first GMO food labeling law. Now it prepares to get sued". The Washington Post. Retrieved 19 January 2016.

7

Food Fortification: Practical Approaches

Food fortification is the addition of nutrients like minerals and vitamins into processed food to help tide over any nutritional deficiencies that could exist or to make food more commercially viable. This chapter contains information on the fortification of food and the common types of fortification. It also has a section on dietary supplements and nutraceuticals. The topics discussed in the chapter are of great importance to broaden the existing knowledge on food engineering.

Food Fortification

Manufacturers once proposed selling fortified junk food and beer, but USFDA policies of the time forbade it

Food fortification or enrichment is the process of adding micronutrients (essential trace elements and vitamins) to food. It may be a purely commercial choice to provide extra nutrients in a food, while other times it is a public health policy which aims to reduce the number of people with dietary deficiencies within a population.

Diets that lack variety can be deficient in certain nutrients. Sometimes the staple foods of a region can lack particular nutrients, due to the soil of the region or because of the inherent inadequacy of the normal diet. Addition of micronutrients to staples and condiments can prevent large-scale deficiency diseases in these cases.

While it is true that both fortification and enrichment refer to the addition of nutrients to food, the true definitions do slightly vary. As defined by the World Health Organization (WHO) and the Food and Agricultural Organization of the United Nations (FAO), fortification refers to "the practice of deliberately increasing the content of an essential micronutrient, ie. vitamins and minerals (including trace elements) in a food irrespective of whether the nutrients were originally in the food before processing or not, so as to improve the nutritional quality of the food supply and to provide a public health benefit with minimal risk to health," whereas enrichment is defined as "synonymous with fortification and refers to the addition of micronutrients to a food which are lost during processing."

Food fortification was identified as the second strategy of four by the WHO and FAO to begin decreasing the incidence of nutrient deficiencies at the global level.

As outlined by the FAO, the most common fortified foods are:

- Cereals and cereal based products
- Milk and Milk products
- Fats and oils
- Accessory food items
- Tea and other beverages
- Infant formulas

Types

The four main methods of food fortification (named as to indicate the procedure that is used in order to fortify the food):

1. Biofortification (i.e. breeding crops to increase their nutritional value, which can include both conventional selective breeding, and modern genetic modification)
2. Synthetic biology (i.e. addition of probiotic bacteria to foods)
3. Commercial and industrial fortification (i.e. flour, rice, oils (common cooking foods))
4. Home fortification (e.g. vitamin D drops)

Rationale

The WHO and FAO, among many other nationally recognized organizations, have recognized that there are over 2 billion people worldwide who suffer from a variety of micronutrient deficiencies. In 1992, 159 countries pledged at the FAO/WHO International Conference on Nutrition to make efforts to help combat these issues of micronutrient deficiencies, highlighting the importance of decreasing the number of those with iodine, vitamin A, and iron deficiencies. A significant statistic that led to these efforts was the discovery that approximately 1 in 3 people worldwide were at risk for either an iodine, vitamin A, or iron deficiency. Although it is recognized that food fortification alone will not combat this deficiency, it is a step towards reducing the prevalence of these deficiencies and their associated health conditions.

In Canada, The Food and Drug Regulations have outlined specific criterion which justifies food fortification:

1. To replace nutrients which were lost during manufacturing of the product (i.e. the manufacturing of flour)
2. To act as a public health intervention
3. To ensure the nutritional equivalence of substitute foods (i.e. to make butter and margarine similar in content, soy milk and cow's milk, etc.)
4. To ensure the appropriate vitamin and mineral nutrient composition of foods for special dietary purposes (i.e.,gluten-free products, low sodium, or any other products specifically designed for special dietary requirements from an individual).

There are also several advantages to approaching nutrient deficiencies among populations via food fortification as opposed to other methods. These may include, but are not limited to: treating a population without specific dietary interventions therefore not requiring a change in dietary patterns, continuous delivery of the nutrient, does not require individual compliance, and potential to maintain nutrient stores more efficiently if consumed on a regular basis.

Criticism

Several organizations such as the WHO, FAO, Health Canada, and the Nestlé Research Center acknowledge that there are limitations to food fortification. Within the discussion of nutrient deficiencies the topic of nutrient toxicities can also be immediately questioned. Fortification of nutrients in foods may deliver toxic amounts of nutrients to an individual and also cause its associated side effects. As seen with the case of excessive fluoride intakes below, the result can be irreversible staining to the teeth. Although this may be a minor toxic effect to health, there are several that are more severe.

The WHO states that limitations to food fortification may include: human rights issues indicating that consumers have the right to choose if they want fortified products or not, the potential for insufficient demand of the fortified product, increased production costs leading to increased retail costs, the potential that the fortified products will still not be a solution to nutrient deficiencies amongst low income populations who may not be able to afford the new product, and children who may not be able to consume adequate amounts thereof.

Food safety worries led to legislation in Denmark in 2004 restricting foods fortified with extra vitamins or minerals. Products banned include: Rice Crispies, Shreddies, Horlicks, Ovaltine and Marmite.

Danes said [Kelloggs] Corn Flakes, Rice Krispies and Special K wanted to include "toxic" doses which, if eaten regularly, could damage children's livers and kidneys and harm fetuses in pregnant women.

Limited Absorption

One factor that limits the benefits of food fortification is that isolated nutrients added back into a processed food that has had many of its nutrients removed, does not always result in the added nutrients being as bioavailable as they would be in the original, whole food. An example is skim milk that has had the fat removed, and then had vitamin A and vitamin D added back. Vitamins A and D are both fat-soluble and non-water-soluble, so a person consuming skim milk in the absence of fats may not be able to absorb as much of these vitamins as one would be able to absorb from drinking whole milk. On the other hand, the nutrient added as a fortificant may have a higher bioavailability than from foods, which is the case with folic acid used to increase folate intakes.

Phytochemicals such as polyphenols, phytic acid in cereal grains and tannin in tea can also impact nutrient absorption, limiting the bioavailability of intrinsic and additional nutrients, and reducing the effectiveness of fortification programs.

Excess Intake of Micronutrients

Ecological studies have shown that increased B vitamin fortification is correlated with the prevalence of obesity and diabetes. Daily consumption of iron per capita in the United States has dramatically surged since World War II and nearly doubled over the past century due to increases in iron fortification and increased consumption of meat. Existing evidence suggests that excess iron intake may play a role in the development of obesity, cardiovascular disease, diabetes and cancer.

Fortification of foods with folic acid has been mandated in many countries solely to improve the folate status of pregnant women to prevent Neural Tube Defects—a relatively rare birth defect which affected 0.5% of US births before fortification began. However, when fortification is introduced, several hundred thousand people are exposed to an increased intake of folic acid for each neural tube defect pregnancy that is prevented. In humans, increased folic acid intake leads to elevated blood concentrations of naturally occurring folates and of unmetabolized folic acid. High blood concentrations of folic acid may decrease natural killer cell cytotoxicity, and high folate status may reduce the response to drugs used to treat malaria, rheumatoid arthritis, psoriasis, and cancer. A combination of high folate levels and low vitamin B-12 status may be associated with an increased risk of cognitive impairment and anemia in the elderly and, in pregnant women, with an increased risk of insulin resistance and obesity in their children. Folate has a dual effect on cancer, protecting against cancer initiation but facilitating progression and growth of preneoplastic cells and subclinical cancers. Furthermore, intake of folic acid from fortification have turned out to be significantly greater than originally modeled in pre mandate predictions. Therefore, a high folic acid intake due to fortification may be harmful for more people than the policy is designed to help.

Different forms of Micronutrients

There is a concern that micronutrients are legally defined in such a way that does not distinguish between different forms, and that fortified foods often have nutrients in a balance that would not occur naturally. For example, in the U.S., food is fortified with folic acid, which is one of the many naturally-occurring forms of folate, and which only contributes a minor amount to the folates occurring in natural foods. In many cases, such as with folate, it is an open question of whether or not there are any benefits or risks to consuming folic acid in this form.

In many cases, the micronutrients added to foods in fortification are synthetic.

In some cases, certain forms of micronutrients can be actively toxic in a sufficiently high dose, even if other forms are safe at the same or much higher doses. There are examples of such toxicity in both synthetic and naturally-occurring forms of vitamins. Retinol, the active form of Vitamin A, is toxic in a much lower dose than other forms, such as beta carotene. Menadione, a phased-out synthetic form of Vitamin K, is also known to be toxic.

Food Supplements

There are several main groups of food supplements like:

- Vitamins and co-vitamins
- Essential minerals
- Essential fatty acids
- Essential amino acids
- Phytonutrients
- Enzymes

Examples of Fortification in Foods

Many foods and beverages worldwide have been fortified, whether a voluntary action by the product developers or by law. Although some may view these additions as strategic marketing schemes to sell their product, there is a lot of work that must go into a product before simply fortifying it. In order to fortify a product, it must first be proven that the addition of this vitamin or mineral is beneficial to health, safe, and an effective method of delivery. The addition must also abide by all food and labeling regulations and support nutritional rationale. From a food developer's point of view, they also need to consider the costs associated with this new product and whether or not there will be a market to support the change.

Examples of foods and beverages that have been fortified and shown to have positive health effects:

Iodized Salt

"Iodine deficiency disorder (IDD) is the single greatest cause of preventable mental retardation. Severe deficiencies cause cretinism, stillbirth and miscarriage. But even mild deficiency can significantly affect the learning ability of populations........ Today over 1 billion people in the world suffer from iodine deficiency, and 38 million babies born every year are not protected from brain damage due to IDD."—Kul Gautam, Deputy Executive Director, UNICEF, October 2007

Iodised salt has been used in the United States since before World War II. It was discovered in 1821 that goiters could be treated by the use of iodized salts. However, it was not until 1916 that the use of iodized salts could be tested in a research trial as a preventative measure against goiters. By 1924, it became readily available in the US.

Currently in Canada and the US, the RDA for iodine is as low as 90 μg/day for children (4–8 years) and as high as 290 μg/day for breast-feeding mothers.

Diseases that are associated with an iodine deficiency include: mental retardation, hypothyroidism, and goiter. There is also a risk of various other growth and developmental abnormalities.

Folic Acid

Folic acid (also known as folate) functions in reducing blood homocysteine levels, forming red blood cells, proper growth and division of cells, and preventing neural tube defects (NTDs).

In many industrialized countries, the addition of folic acid to flour has prevented a significant number of NTDs in infants. Two common types of NTDs, spina bifida and anencephaly, affect approximately 2500-3000 infants born in the US annually. Research trials have shown the ability to reduce the incidence of NTDs by supplementing pregnant mothers with folic acid by 72%.

The RDA for folic acid ranges from as low as 150 μg/day for children aged 1–3 years old, to 400 μg/day for males and females over the age of 19, and 600 μg/day during pregnancy.

Diseases associated with folic acid deficiency include: megaloblastic or macrocytic anemia, cardiovascular disease, certain types of cancer, and NTDs in infants.

Niacin

Niacin has been added to bread in the USA since 1938 (when voluntary addition started), a programme which substantially reduced the incidence of pellagra. As early as 1755, pellagra was recognized by doctors as being a niacin deficiency disease. Although not officially receiving its name of pellagra until 1771. Pellagra was seen amongst poor families who used corn as their main dietary staple. Although corn itself does contain niacin, it is not a bioavailable form unless it undergoes Nixtamalization (treatment with alkali, traditional in Native American cultures) and therefore was not contributing to the overall intake of niacin. Although pellagra can still be seen in developing countries, fortification of food with niacin played a huge role in eliminating the prevalence of the disease.

The RDA for niacin is 2 mg NE(niacin equivalents)/day (AI) for infants aged 0–6 months, 16 mg NE/day for males, and 14 mg NE/day for females who are over the age of 19.

Diseases associated with niacin deficiency include: Pellagra which consisted of signs and symptoms called the 3D's-"Dermatitis, dementia, and diarrhea. Others may include vascular or gastrointestinal diseases.

Common diseases which present a high frequency of niacin deficiency: alcoholism, anorexia nervosa, HIV infection, gastrectomy, malabsorptive disorders, certain cancers and their associated treatments.

Vitamin D

Since Vitamin D is a fat-soluble vitamin, it cannot be added to a wide variety of foods. Foods that it is commonly added to are margarine, vegetable oils and dairy products. During the late 1800s, after the discovery of curing conditions of scurvy and beriberi had occurred, researchers were aiming to see if the disease, later known as rickets, could also be cured by food. Their results showed that sunlight exposure and cod liver oil were the cure. It was not until the 1930s that vitamin D was actually linked to curing rickets. This discovery led to the fortification of common foods such as milk, margarine, and breakfast cereals. This took the astonishing statistics of approximately

80–90% of children showing varying degrees of bone deformations due to vitamin D deficiency to being a very rare condition.

Risk factors for vitamin D deficiencies include:

- In infants, being exclusively or primarily breast-fed
- Dark skin
- Living in cold climates and having little sun exposure
- Being elderly
- Covering all or almost all of one's skin while outdoors
- Liberal use of high-SPF sunscreens
- Fat malabsorption syndromes
- Inflammatory bowel diseases
- Obesity

The current RDA for infants aged 0–6 months is 10 µg (400 International Units (IU))/day and for adults over 19 years of age it is 15 µg (600 IU)/day.

Diseases associated with a vitamin D deficiency include rickets, osteoporosis, and certain types of cancer (breast, prostate, colon and ovaries). It has also been associated with increased risks for fractures, heart disease, type 2 diabetes, autoimmune and infectious diseases, asthma and other wheezing disorders, myocardial infarction, hypertension, congestive heart failure, and peripheral vascular disease.

Fluoride

Although fluoride is not considered an essential mineral, it is seen as possibly useful in prevention of tooth decay and maintaining adequate dental health. In the mid-1900s it was discovered that towns with a high level of fluoride in their water supply was causing the residents' teeth to have both brown spotting and a strange resistance to dental caries. This led to the fortification of water supplies with fluoride with safe amounts to retain the properties of resistance to dental caries but avoid the staining cause by fluorosis (a condition caused by excessive fluoride intake). The tolerable upper intake level (UL) set for fluoride ranges from 0.7 mg/day for infants aged 0–6 months and 10 mg/day for adults over the age of 19.

Conditions commonly associated with fluoride deficiency are dental caries and osteoporosis.

Others

Some other examples of fortified foods:

- Calcium is frequently added to fruit juices, carbonated beverages and rice.
- White rice is frequently enriched to replace some of the lost nutrients during milling or adding extras in.

- "Golden rice" is a variety of rice which has been genetically modified to produce beta carotene.
- Amylase rich flour is utilized for food making to increase dietary consumption.

Fortification for Body Building

Despite having some scientific basis, but with controversial ethics, is the science of using foods and food supplements to achieve a defined health goal. A common example of this use of food supplements is the extent to which body builders will use amino acid mixtures, vitamins and phytochemicals to enhance natural hormone production, increase muscle and reduce fat. The literature is not concrete on an appropriate method for use of fortification for body builders and therefore may not be recommended due to safety concerns.

Fortification for Medical Treatment

There is interest in the use of food supplements in established medical conditions. This nutritional supplementation using foods as medicine (nutraceuticals) has been effectively used in treating disorders affecting the immune system up to and including cancers. This goes beyond the definition of "food supplement", but should be included for the sake of completeness.

Example of Fortification of Food

Folic acid

Folic acid (conjugate base folate) is a B vitamin. Folic acid is made and used in fortified foods and supplements on the theory that it is converted into folate. However, folic acid is an oxidized form, not significantly found in fresh natural foods. To be used it must be converted to tetrahydrofolate (tetrahydrofolic acid) by dihydrofolate reductase (DHFR). Increasing evidence suggests that this process may be slow in humans. Folate and folic acid derive their names from the Latin word *folium*, which means "leaf". Folates occur naturally in many foods and, among plants, are especially plentiful in dark green leafy vegetables.

Vitamin B_9 is essential for numerous bodily functions. Humans cannot make folates; therefore, folic acid has to be supplied through the diet to meet their daily requirements. The human body needs folate to make DNA, repair DNA, and methylate DNA as well as to act as a cofactor in certain biological reactions. It is especially important in aiding rapid cell division and growth, such as in infancy and pregnancy. Children and adults both require folate to produce healthy red blood cells and prevent anemia.

A lack of dietary folates can lead to folate deficiency. A complete lack of dietary folate takes months before deficiency develops as normal individuals have about 500–20,000 micrograms (μg) of folate in body stores. This deficiency can result in health problems. During pregnancy deficiency increases the risk of neural tube defects in the baby—a birth defect affecting 300,000 (0.2%) births globally each year and 3,000 pregnancies in the United States each year. Common symptoms of folate deficiency include diarrhea, macrocytic anemia with weakness or shortness of breath, nerve damage with weakness and limb numbness, pregnancy complications, mental confusion, forgetfulness or other cognitive deficits, mental depression, sore or swollen tongue, peptic or mouth ulcers, headaches, heart palpitations, irritability, and behavioral disorders. Low levels of folate can also lead to homocysteine accumulation. Low levels of folate have been associated with specific cancers. However, it is not clear whether consuming recommended (or higher) amounts of folic acid—from foods or in supplements—can lower cancer risk.

Definition

Food supplement manufacturers often use the term *folate* for something different from "pure" folic acid: in chemistry, *folate* refers to the deprotonated ion, and *folic acid* to the neutral molecule—which both coexist in water. The International Union of Pure and Applied Chemistry and the International Union of Biochemistry and Molecular Biology state that folate and folic acid are the preferred synonyms for pteroylglutamate and pteroylglutamic acid, respectively. *Folate* indicates a collection of "folates" that is not chemically well-characterized, including other members of the family of pteroylglutamates, or mixtures of them, having various levels of reduction of the pteridine ring, one-carbon substitutions and different numbers of glutamate residues.

Health Effects

Pregnancy

Folic acid intake during pregnancy has been linked to a lessened risk of neural tube defects and some other specific kinds of birth defects.

There is growing concern worldwide that prenatal high folic acid in the presence of low vitamin B12 causes epigenetic changes in the unborn predisposing them to metabolic syndromes, central adiposity and adult diseases such as Type 2 diabetes.

Despite concerns, there appears to be no correlation between maternal folic acid supplementation and an increased risk for asthma in the child

Fertility

Folate is necessary for fertility in both men and women. It contributes to spermatogenesis. Therefore, it is necessary to receive sufficient amounts through the diet to avoid subfertility. Also, polymorphisms in genes of enzymes involved in folate metabolism could be one reason for fertility complications in some women with unexplained infertility.

Heart Disease

Taking folic acid does not reduce cardiovascular disease even though it reduces homocysteine levels.

Folic acid supplements consumed before and during pregnancy may reduce the risk of heart defects in infants.

Stroke

Folic acid appears to reduce the risk of stroke, which may be due to the role folate plays in regulating homocysteine concentration. The reviews indicate the risk of stroke appears to be reduced only in some individuals, but a definite recommendation regarding supplementation beyond the current RDA has not been established for stroke prevention. Observed stroke reduction is consistent with the reduction in pulse pressure produced by folate supplementation of 5 mg per day, since hypertension is a key risk factor for stroke. Folic supplements are inexpensive and relatively safe to use, which is why stroke or hyperhomocysteinemia patients are encouraged to consume daily B vitamins including folic acid.

Cancer

Folic acid supplementation may slightly increase the risk of some cancers.

Antifolate Chemotherapy

Folate is important for cells and tissues that rapidly divide. Cancer cells divide rapidly, and drugs that interfere with folate metabolism are used to treat cancer. The antifolate methotrexate is a drug often used to treat cancer because it inhibits the production of the active form of THF from the inactive dihydrofolate (DHF). However, methotrexate can be toxic, producing side effects, such as inflammation in the digestive tract that make it difficult to eat normally. Also, bone marrow depression (inducing leukopenia and thrombocytopenia), and acute kidney and liver failure have been reported.

Folinic acid, under the drug name leucovorin, a form of folate (formyl-THF), can help "rescue" or reverse the toxic effects of methotrexate. Folinic acid is *not* the same as folic acid. Folic acid supplements have little established role in cancer chemotherapy. There have been cases of severe adverse effects of accidental substitution of folic acid for folinic acid in patients receiving methotrexate cancer chemotherapy. It is important for anyone receiving methotrexate to follow medical advice on the use of folic or folinic acid supplements. The supplement of folinic acid in patients undergoing methotrexate treatment is to give cells dividing less rapidly enough folate to maintain normal cell functions. The amount of folate given is depleted by rapidly dividing cells (cancer) quickly, and so does not negate the effects of methotrexate.

Psychological

Some evidence links a shortage of folate with depression. Limited evidence from randomised controlled trials showed using folic acid in addition to SSRIs may have benefits. Research at the University of York and Hull York Medical School has found a link between depression and low levels of folate. One study by the same team involved 15,315 subjects. However, the evidence is probably too limited at present for this to be a routine treatment recommendation.Folic acid supplementation affects noradrenaline and serotonin receptors within the brain, which could be the cause of folic acid's possible ability to act as an antidepressant. The exact mechanisms involved in the development of schizophrenia and depression are not entirely clear, but the bioactive folate, methyltetrahydrofolate (5-MTHF), a direct target of methyl donors like S-adenosyl methionine (SAMe), recycles the inactive dihydrobiopterin (BH_2) into tetrahydrobiopterin (BH_4), the necessary cofactor in various steps of monoamine synthesis, including that of dopamine. BH_4 serves a regulatory role in monoamine neurotransmission and is required to mediate the actions of most antidepressants. 5-MTHF also plays both direct & indirect roles in DNA methylation, NO_2 synthesis, and one-carbon metabolism.

Age Related Macular Degeneration

A substudy of the Women's Antioxidant and Folic Acid Cardiovascular Study published in 2009 reported use of a nutritional supplement containing folic acid, pyridoxine, and cyanocobalamin decreased the risk of developing age-related macular degeneration by 34.7%.

Folic Acid, B_{12} and Iron

There is a complex interaction between folic acid, vitamin B_{12} and iron. A deficiency of one may be "masked" by excess of another so the three must always be in balance.

Toxicity

The risk of toxicity from folic acid is low, because folate is a water-soluble vitamin and is regularly removed from the body through urine. One potential issue associated with high dosages of folic acid is that it has a masking effect on the diagnosis of pernicious anaemia (vitamin B_{12} deficiency), and a variety of concerns of potential negative impacts on health.

Folate Deficiency

Folate deficiency can be caused by unhealthy diets that do not include enough fruits and vegetables, diseases in which folic acid is not well absorbed in the digestive system (such as Crohn's disease or Celiac disease), some genetic disorders that affect levels of folate, and certain medicines (such as phenytoin, sulfasalazine, or trimethoprim-sulfamethoxazole). Folate deficiency is accelerated by alcohol consumption.

Folate deficiency may lead to glossitis, diarrhea, depression, confusion, anemia, and fetal neural tube defects and brain defects (during pregnancy). Other symptoms include fatigue, gray hair, mouth sores, poor growth, and swollen tongue. Folate deficiency is diagnosed by analyzing CBC and plasma vitamin B_{12} and folate levels. CBC may indicate megaloblastic anemia but this could also be a sign of vitamin B_{12} deficiency. A serum folate of 3 µg/L or lower indicates deficiency. Serum folate level reflects folate status but erythrocyte folate level better reflects tissue stores after intake. Serum folate reacts more rapidly to folate intake than erythrocyte folate. An erythrocyte folate level of 140 µg/L or lower indicates inadequate folate status. Increased homocysteine level suggests tissue folate deficiency but homocysteine is also affected by vitamin B_{12} and vitamin B_6, renal function, and genetics.

One way to differentiate between folate deficiency from vitamin B_{12} deficiency is by testing for methylmalonic acid levels. Normal MMA levels indicate folate deficiency and elevated MMA levels indicate vitamin B_{12} deficiency. Folate deficiency is treated with supplemental oral folate of 400 to 1000 µg per day. This treatment is very successful in replenishing tissues, even if deficiency was caused by malabsorption. Patients with megaloblastic anemia need to be tested for vitamin B_{12} deficiency before folate treatment, because if the patient has vitamin B_{12} deficiency, folate supplementation can remove the anemia, but can also worsen neurologic problems. Morbidly obese patients with BMIs of greater than 50 are more likely to develop folate deficiency. Patients with celiac disease have a higher chance of developing folate deficiency. Cobalamin deficiency may lead to folate deficiency, which, in turn, increases homocysteine levels and may result in the development of cardiovascular disease or birth defects.

Malaria

Some studies show iron-folic acid supplementation in children under 5 may result in increased mortality due to malaria; this has prompted the World Health Organization to alter their iron-folic acid supplementation policies for children in malaria-prone areas, such as India.

Dietary Reference Intake

Because of the difference in bioavailability between supplemented folic acid and the different forms of folate found in food, the dietary folate equivalent (DFE) system was established. One DFE is defined as 1 µg of dietary folate, or 0.6 µg of folic acid supplement.

National Institutes of Health (US) Nutritional Requirements (µg per day)								
Age	Infants (AI)	Infants (UL)	Adults (RDA)	Adults (UL)	Pregnant women (RDA)	Pregnant women (UL)	Lactating women (RDA)	Lactating women (UL)
0–6 months	65	None set	–	–	–	–	–	–
7–12 months	80	None set	–	–	–	–	–	–
1–3 years	–	–	150	300	–	–	–	–
4–8 years	–	–	200	400	–	–	–	–
9–13 years	–	–	300	600	–	–	–	–
14–18	–	–	400	800	600	800	500	800
19+	–	–	400	1000	600	1000	500	1000

The Dietary Reference Intakes (DRIs) were developed by the United States National Academy of Sciences to set reference values for planning and assessing nutrient intake for healthy people. DRIs incorporate four reference values: Estimated Average Requirements (EARs), Recommended Dietary Allowances (RDAs; the daily intake level that is adequate for 97–98% of the population in the United States), Adequate Intakes, for when the existing science is inadequate to set EARs and RDAs, and tolerable upper intake levels (UL, the highest level of intake considered safe). The UL for folate refers to only synthetic folic acid, as no health risks have been associated with high intake of folate from food sources. The European Food Safety Authority reviewed the same safety question and also set the adult UL at 1000 µg.

For U.S. food and dietary supplement labeling purposes the amount in a serving is expressed as a percent of Daily Value (%DV). For folic acid labeling purposes 100% of the Daily Value was 400 µg. As of the May 2016 update it was kept unchanged at 400 µg. A table of the pre-change adult Daily Values is provided at Reference Daily Intake. Food and supplement companies have until July 28, 2018 to comply with the change.

Sources

Vitamins C and M as featured on a monument in front of University of Warsaw's Centre of New Technologies

Folate naturally occurs in a wide variety of foods, including vegetables (particularly dark green leafy vegetables), fruits and fruit juices, nuts, beans, peas, dairy products, poultry and meat, eggs, seafood, grains, and some beers. Avocado, beetroot, spinach, liver, yeast, asparagus, and Brussels sprouts are among the foods with the highest levels of folate.

Folic acid is added to grain products in many countries, and in these countries, fortified products make up a significant source of the population's folic acid intake. Because of the difference in bioavailability between supplemented folic acid and the different forms of folate found in food, the dietary folate equivalent (DFE) system was established. 1 DFE is defined as 1 µg of dietary folate, or 0.6 µg of folic acid supplement. This is reduced to 0.5 µg of folic acid if the supplement is taken on an empty stomach.

Folate naturally found in food is susceptible to high heat and ultraviolet light, and is soluble in water. It is heat-labile in acidic environments and may also be subject to oxidation.

Some meal replacement products do not meet the folate requirements as specified by the RDAs.

Folate (B9) can also be processed from the provitamin Pteroylmonoglutamic acid (Vitamin B10).

History

In the 1920s, scientists believed folate deficiency and anemia were the same condition. In 1931, researcher Lucy Wills made a key observation that led to the identification of folate as the nutrient required to prevent anemia during pregnancy. Dr. Wills demonstrated that anemia could be reversed with brewer's yeast. In the late 1930s, folate was identified as the corrective substance in brewer's yeast.

It was first isolated in and extracted from spinach leaves by Mitchell and others in 1941. Bob Stokstad isolated the pure crystalline form in 1943, and was able to determine its chemical structure while working at the Lederle Laboratories of the American Cyanamid Company. This historical research project, of obtaining folic acid in a pure crystalline form in 1945, was done by the team called the "folic acid boys," under the supervision and guidance of Director of Research Dr. Yellapragada Subbarow, at the Lederle Lab, Pearl River, NY.

This research subsequently led to the synthesis of the antifolate aminopterin, the first-ever anticancer drug, the clinical efficacy was proven by Sidney Farber in 1948. In the 1950s and 1960s, scientists began to discover the biochemical mechanisms of action for folate. In 1960, experts first linked folate deficiency to neural tube defects. In the late 1990s, US scientists realized, despite the availability of folate in foods and in supplements, there was still a challenge for people to meet their daily folate requirements, which is when the US implemented the folate fortification program.

Biological Roles

DNA and Cell Division

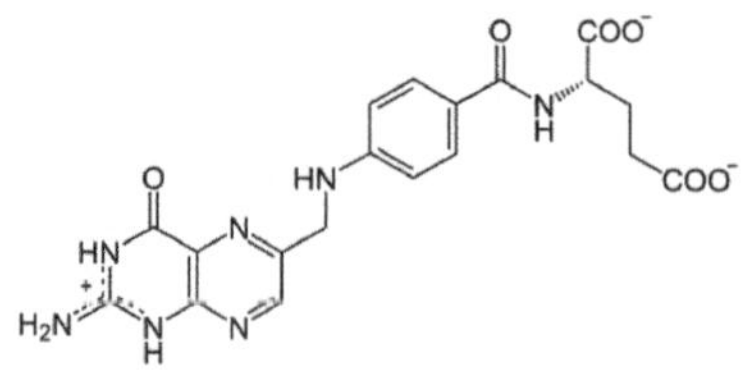

A diagram of the chemical structure of folate

Folate is necessary for the production and maintenance of new cells, for DNA synthesis and RNA synthesis, and for preventing changes to DNA, and, thus, for preventing cancer. It is especially important during periods of frequent cell division and growth, such as infancy and pregnancy. Folate is needed to carry one-carbon groups for methylation reactions and nucleic acid synthesis (the most notable one being thymine, but also purine bases). Thus, folate deficiency hinders DNA synthesis and cell division, affecting hematopoietic cells and neoplasms the most because of their greater frequency of cell division. RNA transcription, and subsequent protein synthesis, are less affected by folate deficiency, as the mRNA can be recycled and used again (as opposed to DNA synthesis, where a new genomic copy must be created). Since folate deficiency limits cell division, erythropoiesis, production of red blood cells, is hindered and leads to megaloblastic anemia, which is characterized by large immature red blood cells. This pathology results from persistently thwarted attempts at normal DNA replication, DNA repair, and cell division, and produces abnormally large red cells called megaloblasts (and hypersegmented neutrophils) with abundant cytoplasm capable of RNA and protein synthesis, but with clumping and fragmentation of nuclear chromatin. Some of these large cells, although immature (reticulocytes), are released early from the marrow in an attempt to compensate for the anemia. Both adults and children need folate to make normal red and white blood cells and prevent anemia. Deficiency of folate in pregnant women has been implicated in neural tube defects (NTD); therefore, many developed countries have implemented mandatory folic acid fortification in cereals, etc. NTDs occur early in pregnancy (first month), therefore women must have abundant folate upon conception. Folate is required to make red blood cells and white blood cells and folate deficiency may lead to anemia, which causes fatigue, weakness and inability to concentrate.

DNA and Amino Acid Production

In the form of a series of tetrahydrofolate (THF) compounds, folate derivatives are substrates in a number of single-carbon-transfer reactions, and also are involved in the synthesis of dTMP (2′-deoxythymidine-5′-phosphate) from dUMP (2′-deoxyuridine-5′-phosphate). It is a substrate for an important reaction that involves vitamin B_{12} and it is necessary for the synthesis of DNA, and so required for all dividing cells.

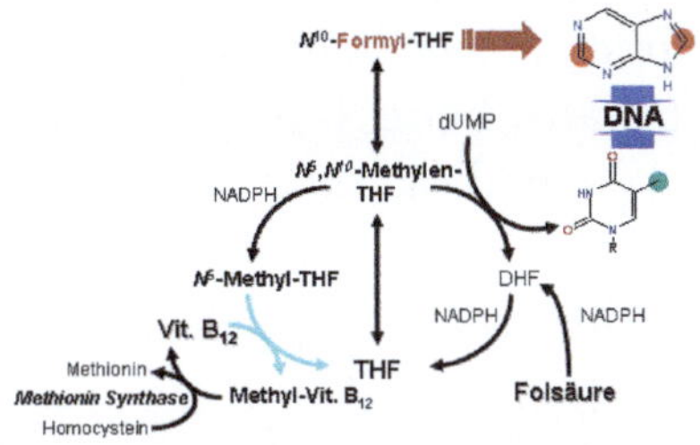

Metabolism of folic acid to recycle homocysteine into methionine

The pathway leading to the formation of tetrahydrofolate (FH_4) begins when folic acid (F) is reduced to dihydrofolate (DHF) (FH_2), which is then reduced to THF. Dihydrofolate reductase catalyses the last step. Vitamin B_3 in the form of NADPH is a necessary cofactor for both steps of the synthesis. Thus, hydride molecules are transferred from NADPH to the C6 position of the pteridine ring to reduce folic acid to THF.

Methylene-THF (CH_2FH_4) is formed from THF by the addition of a methylene bridge from one of three carbon donors: formate, serine, or glycine. Methyl tetrahydrofolate (CH_3-THF, or methyl-THF) can be made from methylene-THF by reduction of the methylene group with NADPH.

Another form of THF, 10-formyl-THF, results from oxidation of methylene-THF or is formed from formate donating formyl group to THF. Also, histidine can donate a single carbon to THF to form methenyl-THF.

Vitamin B_{12} is the only acceptor of methyl-THF, and this reaction produces methyl-B_{12} (methylcobalamin). There is also only one acceptor for methyl-B_{12}, homocysteine, in a reaction catalyzed by homocysteine methyltransferase. These reactions are important because a defect in homocysteine methyltransferase or a deficiency of B_{12} may lead to a so-called "methyl-trap" of THF, in which THF converts to a reservoir of methyl-THF. Thereafter, this THE has no way of being metabolized, and serves as a sink of THF that causes a subsequent deficiency in folate. Thus, a deficiency in B_{12} can generate a large pool of methyl-THF that is unable to undergo reactions and mimics folate deficiency.

The reactions that lead to the methyl-THF reservoir can be shown in chain form:

folate → dihydrofolate → tetrahydrofolate ↔ methylene-THF → methyl-THF

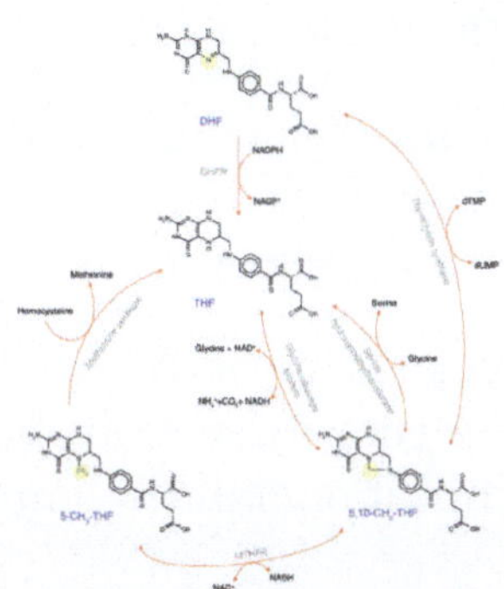

Folate metabolism

Conversion to Biologically Active Derivatives

All the biological functions of folic acid are performed by tetrahydrofolate and other derivatives. Their biological availability to the body depends upon dihydrofolate reductase action in the liver. This action is unusually slow in humans, being less than 2% of that in rats (and with an almost-5-fold variation in enzymatic activity), leading to the accumulation of unmetabolized folic acid. It has been suggested this low activity limits the conversion of folic acid into its biologically active forms "when folic acid is consumed at levels higher than the Tolerable Upper Intake Level (1 mg/d for adults)."

Overview of Drugs that Interfere with Folate Reactions

A number of drugs interfere with the biosynthesis of folic acid and THF. Among them are the dihydrofolate reductase inhibitors such as trimethoprim, pyrimethamine, and methotrexate; the sulfonamides (competitive inhibitors of 4-aminobenzoic acid in the reactions of dihydropteroate synthetase).

Valproic acid, one of the most commonly prescribed anticonvulsants that is also used to treat certain psychological conditions, is a known inhibitor of folic acid, and as such, has been shown to cause neural tube defects and cases of spina bifida and cognitive impairment in the newborn. Because of this considerable risk, those mothers who must continue to use valproic acid or its derivatives during pregnancy to control their condition (as opposed to stopping the drug or switching to another drug or to a lesser dose) should take folic acid supplements under the direction and guidance of their health care providers.

The National Health and Nutrition Examination Survey (NHANES III 1988–91) and the Continuing Survey of Food Intakes by Individuals (1994–96 CSFII) indicated most adults did not consume adequate folate. However, the folic acid fortification program in the United States has increased folic acid content of commonly eaten foods such as cereals and grains, and as a result, diets of most adults now provide recommended amounts of folate equivalents.

Food Fortification

Folic acid fortification is a process where folic acid is added to flour with the intention of promoting public health through increasing blood folate levels in the populace. In the USA, food is fortified with folic acid, only one of the many naturally-occurring forms of folate, and a substance contributing only a minor amount to the folates in natural foods.

Since the discovery of the link between insufficient folic acid and neural tube defects, governments and health organizations worldwide have made recommendations concerning folic acid *supplementation* for women intending to become pregnant.

Fortification is controversial, with issues having been raised concerning individual liberty, as well as the health concerns described in the Toxicity section above. In the USA, there is concern that the federal government mandates fortification, but does not provide monitoring of potential undesirable effects of fortification.

76 countries worldwide require mandatory folic acid fortification of at least one major cereal grain,

with nearly all fortifying at least wheat flour, according to November 2013 data from the Flour Fortification Initiative. These countries are:

> Antigua and Barbuda, Argentina, Australia, Bahamas, Bahrain, Barbados, Belize, Benin, Bolivia (Plurinational State of), Brazil, Burkina Faso, Cameroon, Canada, Chile, Colombia, Costa Rica, Cote d'Ivoire, Cuba, Dominica, Dominican Republic, Ecuador, Egypt, El Salvador, Fiji, Ghana, Grenada, Guatemala, Guinea, Guyana, Haiti, Honduras, Indonesia, Iran (Islamic Republic of), Iraq, Jamaica, Jordan, Kazakhstan, Kenya, Kosovo, Kuwait, Kyrgyzstan, Liberia, Mali, Mauritania, Mexico, Morocco, Nepal, Nicaragua, Niger, Nigeria, Oman, Palestine (Occupied Territory), Panama, Papua New Guinea, Paraguay, Peru, Republic of Moldova, Rwanda, Saint Kitts and Nevis, Saint Lucia, Saint Vincent and the Grenadines, Saudi Arabia, Senegal, Sierra Leone, Solomon Islands, South Africa, Suriname, Tanzania (United Republic of), Togo, Trinidad and Tobago, Turkmenistan, Uganda, United States of America, Uruguay, Uzbekistan, and Yemen.

As of November 2013, no EU country has mandated folic acid fortification.

Australia

There has been previous debate in Australia regarding the inclusion of folic acid in products such as bread and flour.

Australia and New Zealand have jointly agreed to fortification though the Food Standards Australia New Zealand. Australia will fortify all flour from 18 September 2009. Although the food standard covers both Australia and New Zealand, an Australian government official has stated it is up to New Zealand to decide whether to implement it there, and they will watch with interest.

The requirement is 0.135 mg of folate per 100g of bread.

Canada

In 2003, a Hospital for Sick Children, University of Toronto research group published findings showing the fortification of flour with folic acid in Canada has resulted in a dramatic decrease in neuroblastoma, an early and very dangerous cancer in young children. In 2009, further evidence from McGill University showed a 6.2% decrease per year in the birth prevalence of severe congenital heart defects.

Folic acid used in fortified foods is a synthetic form called pteroylmonoglutamate. It is in its oxidized state and contains only one conjugated glutamate residue. Folic acid therefore enters via a different carrier system from naturally occurring folate, and this may have different effects on folate binding proteins and its transporters. Folic acid has a higher bioavailability than natural folates and are rapidly absorbed across the intestine, therefore it is important to consider the Dietary Folate Equivalent (DFE) when calculating one's intake. Natural occurring folate is equal to 1 DFE, however 0.6 μg of folic acid is equal to 1 DFE.

Folic acid food fortification became mandatory in Canada in 1998, with the fortification of 150 μg of folic acid per 100 grams of enriched flour and uncooked cereal grains. The purpose of fortification was to decrease the risk of neural tube defects in newborns. It is important to fortify grains because it is a widely eaten food and the neural tube closes in the first four weeks of gestation, often before many women even know they are pregnant. Canada's fortification program has been successful with a decrease of neural tube defects by 19% since its introduction. A seven-province study from 1993 to 2002 showed a reduction of 46% in the overall rate of neural tube defects after folic acid fortification was introduced in Canada. The fortification program was estimated to raise a person's folic acid intake level by 70–130 µg/day, however an increase of almost double that amount was actually observed. This could be from the fact that many foods are over fortified by 160–175% the predicted value. In addition, much of the elder population take supplements that adds 400 µg to their daily folic acid intake. This is a concern because 70–80% of the population have detectable levels of unmetabolized folic acid in their blood and high intakes can accelerate the growth of preneoplasmic lesions. It is still unknown the amount of folic acid supplementation that might cause harm.

Supplementation Promotion

According to a Canadian survey, 58% of women said they took a folic acid containing multivitamin or a folic acid supplement as early as three months before becoming pregnant. Women in higher income households and with more years of school education are using more folic acid supplements before pregnancy. Women with planned pregnancies and who are over the age of 25 are more likely to use folic acid supplement. Canadian public health efforts are focused on promoting awareness of the importance of folic acid supplementation for all women of childbearing age and decreasing socio-economic inequalities by providing practical folic acid support to vulnerable groups of women.

New Zealand

New Zealand was planning to fortify bread (excluding organic and unleavened varieties) from 18 September 2009, but has opted to wait until more research is done.

The Association of Bakers and the Green Party have opposed mandatory fortification, describing it as "mass medication". Food Safety Minister Kate Wilkinson reviewed the decision to fortify in July 2009, citing links between overconsumption of folate with cancer . The New Zealand Government is reviewing whether it will continue with the mandatory introduction of folic acid to bread.

United Kingdom

There has been previous debate in the United Kingdom regarding the inclusion of folic acid in products such as bread and flour. While the Food Standards Agency has recommended folic acid fortification, and wheat flour is fortified with iron, folic acid fortification of wheat flour is allowed voluntarily rather than required.

United States

The United States Public Health Service recommends an extra 0.4 mg/day for newly pregnant women, which they can take as a pill. However, many researchers believe this supplementation can never work effectively enough, since about half of all pregnancies in the U.S. are unplanned, and not all women comply with the recommendation. Approximately 53% of the US population uses dietary supplements and 35% uses dietary supplements that contain folic acid.

Men consume more folate (in dietary folate equivalents) than women, and non-Hispanic whites have higher folate intakes than Mexican Americans and non-Hispanic blacks. Twenty-nine percent of black women have inadequate intakes of folate. The age group consuming the most folate and folic acid is the >50 group. 5% of the population exceeds the Tolerable Upper Intake Level.

In 1996, the United States Food and Drug Administration (FDA) published regulations requiring the addition of folic acid to enriched breads, cereals, flours, corn meals, pastas, rice, and other grain products. This ruling took effect on January 1, 1998, and was specifically targeted to reduce the risk of neural tube birth defects in newborns. There are concerns that the amount of folate added is insufficient . In October 2006, the Australian press claimed that U.S. regulations requiring fortification of grain products were being interpreted as disallowing fortification in non-grain products, specifically Vegemite (an Australian yeast extract containing folate). The FDA later said the report was inaccurate, and no ban or other action was being taken against Vegemite.

As a result of the folic acid fortification program, fortified foods have become a major source of folic acid in the American diet. The Centers for Disease Control and Prevention in Atlanta, Georgia used data from 23 birth defect registries covering about half of United States births, and extrapolated their findings to the rest of the country. These data indicate that, since the addition of folic acid in grain-based foods as mandated by the FDA, the rate of neural tube defects dropped by 25% in the United States. Before folic acid fortification, about 4,100 pregnancies were affected by a neural tube defect each year in the United States. After fortification, this number declined to around 3,000. The results of folic acid fortification on the rate of neural tube defects in Canada have also been positive, showing a 46% reduction in prevalence of NTDs; the magnitude of reduction was proportional to the prefortification rate of NTDs, essentially removing geographical variations in rates of NTDs seen in Canada before fortification.

When the U.S. Food and Drug Administration set the folic acid fortification regulation in 1996, the projected increase in folic acid intake was 100 μg/d. Data from a study with 1480 subjects showed that folic acid intake increased by 190 μg/d and total folate intake increased by 323 μg dietary folate equivalents (DFE)/d. Folic acid intake above the upper tolerable intake level (1000 μg folic acid/d) increased only among those individuals consuming folic acid supplements as well as folic acid found in fortified grain products. Taken together, folic acid fortification has led to a bigger increase in folic acid intake than first projected.

Niacin

Niacin, also known as vitamin B_3 and nicotinic acid, is an organic compound with the formula $C_6H_5NO_2$ and, depending on the definition used, one of the 20 to 80 essential human nutrients. Pharmaceutical and supplemental niacin are primarily used to treat hypercholesterolemia (high cholesterol) and pellagra (niacin deficiency). Insufficient niacin in the diet can cause nausea, skin and mouth lesions, anemia, headaches, and tiredness. The lack of niacin may also be observed in pandemic deficiency disease, which is caused by a lack of five crucial vitamins (niacin, vitamin C, thiamin, vitamin D, and vitamin A) and is usually found in areas of widespread poverty and malnutrition. Niacin has not been found to be useful in decreasing the risk of cardiovascular disease in those already on a statin but appears to be effective in those not taking a statin.

This colorless, water-soluble solid is a derivative of pyridine, with a carboxyl group (COOH) at the 3-position. Other forms of vitamin B_3 include the corresponding amide and nicotinamide ("niacinamide"), where the carboxyl group has been replaced by a carboxamide group (CONH 2), as well as more complex amides and a variety of esters. Nicotinic acid and niacinamide are convertible to each other with steady world demand rising from 8,500 tonnes per year in the 1980s to 40,000 in recent years.

Niacin cannot be directly converted to nicotinamide, but both compounds are precursors of the coenzymes nicotinamide adenine dinucleotide (NAD) and nicotinamide adenine dinucleotide phosphate (NADP) *in vivo*. NAD converts to NADP by phosphorylation in the presence of the enzyme NAD+ kinase. NADP and NAD are coenzymes for many dehydrogenases, participating in many hydrogen transfer processes. NAD is important in catabolism of fat, carbohydrate, protein, and alcohol, as well as cell signaling and DNA repair, and NADP mostly in anabolism reactions such as fatty acid and cholesterol synthesis. High energy requirements (brain) or high turnover rate (gut, skin) organs are usually the most susceptible to their deficiency. Although the two are identical in their vitamin activity, nicotinamide does not have the same pharmacological effects (lipid modifying effects) as niacin. Nicotinamide does not reduce cholesterol or cause flushing. As the precursor for NAD and NADP, niacin is also involved in DNA repair.

Dietary Reference Intake

The Food and Nutrition Board of the U.S. Institute of Medicine updated Estimated Average Requirements (EARs) and Recommended Dietary Allowances (RDAs) for niacin in 1998. The current EARs for niacin for women and men ages 14 and up are 11 mg/day and 12 mg/day, respectively; the RDAs are 14 and 16 mg/day, respectively. RDAs are higher than EARs so as to identify amounts that will cover people with higher than average requirements. RDA for pregnancy equals 18 mg/day. RDA for lactation equals 17 mg/day. For infants up to 12 months the Adequate Intake (AI) is 2–4 mg/day. and for children ages 1–13 years the RDA increases with age from 6 to 12 mg/day. As for safety, the Food and Nutrition Board also sets Tolerable Upper Intake Levels (known as ULs) for vitamins and minerals when evidence is sufficient. In the case of niacin the UL is set at 35 mg/day. The European Food Safety Authority reviewed the same safety question and set its UL at 10 mg/day. Safety issues are presented at length in the Side Effects section. Collectively the EARs, RDAs, AIs and ULs are referred to as Dietary Reference Intakes.

For U.S. food and dietary supplement labeling purposes the amount in a serving is expressed as a percent of Daily Value (%DV). For niacin labeling purposes 100% of the Daily Value was 20 mg, but as of May 2016 it has been revised to 16 mg. A table of the pre-change adult Daily Values is provided at Reference Daily Intake. Food and supplement companies have until July 28, 2018 to comply with the change.

Deficiency

Between 1906 and 1940 more than 3 million Americans were affected by pellagra with more than 100,000 deaths. Dr. Joseph Goldberger was assigned to study pellagra by the Surgeon General of the United States and produced good results. In the late 1930s, studies by Dr. Tom Spies, Marion Blankenhorn, and Clark Cooper established that niacin cured pellagra in humans. The disease was greatly reduced as a result.

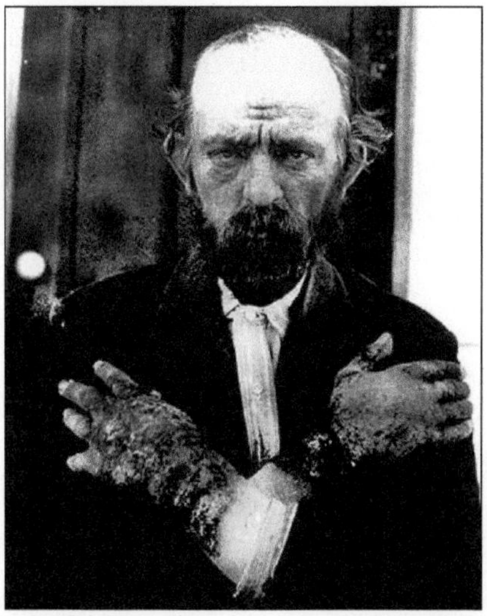

A man with pellagra, which is caused by a chronic lack of vitamin B_3 in the diet

At present, niacin deficiency is sometimes seen in developed countries, and it is usually apparent in conditions of poverty, malnutrition, and chronic alcoholism. It also tends to occur in areas where people eat maize (corn) as a staple food, since it is the only grain low in digestible niacin. A cooking technique called nixtamalization increases the bioavailability of niacin during maize meal/flour production.

Mild niacin deficiency has been shown to slow metabolism, causing decreased tolerance to cold.

Severe deficiency of niacin in the diet causes the disease pellagra, which is characterized by diarrhea, dermatitis, and dementia, as well as Casal's necklace lesions on the lower neck, hyperpigmentation, thickening of the skin, inflammation of the mouth and tongue, digestive disturbances, amnesia, delirium, and eventually death, if left untreated. Common psychiatric symptoms of niacin deficiency include irritability, poor concentration, anxiety, fatigue, restlessness, apathy, and depression. Studies have indicated that, in patients with alcoholic pellagra, niacin deficiency may be an important factor influencing both the onset and severity of this condition. Patients with alcoholism typically experience increased intestinal permeability, leading to negative health outcomes.

Hartnup disease is a hereditary nutritional disorder resulting in niacin deficiency. This condition was first identified in the 1950s by the Hartnup family in London. It is due to a deficit in the intestines and kidneys, making it difficult for the body to break down and absorb dietary tryptophan. The resulting condition is similar to pellagra, including symptoms of red, scaly rash, and sensitivity to sunlight. Oral niacin is given as a treatment for this condition in doses ranging from 40–200 mg, with a good prognosis if identified and treated early. Niacin synthesis is also deficient in carcinoid syndrome, because of metabolic diversion of its precursor tryptophan to form serotonin.

Medical Uses

Abnormal lipids

Niacin is sometimes used in addition to other lipid-lowering medication.

A 2014 review of niacin did not find that it affected either cardiovascular disease or risk of death in those already taking a statin, in spite of raising HDL cholesterol. Niacin alone appears to reduce the rate of cardiovascular events, but not the rate of coronary or cardiovascular deaths.

Treatment of Deficiency

Nicotinic acid has been also approved by the FDA for pellagra.

Contraindications

Niacin is contraindicated with active liver disease, persistent elevated serum transaminases, active peptic ulcer disease, or arterial bleeding.

Side Effects

The most common adverse effects are flushing (e.g., warmth, redness, itching and/or tingling), headache, pain, abdominal pain, diarrhea, dyspepsia, nausea, vomiting, rhinitis, pruritus and rash. These can be minimized by initiating therapy at low dosages, increasing dosage gradually, and avoiding administration on an empty stomach. High doses of niacin often temporarily reduce blood pressure as a result of acute vasodilation. Longer term, high-dose niacin use may persistently lower blood pressure in individuals with hypertension, but more research is needed to determine the extent of this effect.

Facial Flushing

Flushing usually lasts for about 15 to 30 minutes, though it can sometimes last up to two hours. It is sometimes accompanied by a prickly or itching sensation, in particular, in areas covered by clothing. Flushing can be blocked by taking 300 mg of aspirin half an hour before taking niacin, by taking one tablet of ibuprofen per day or by co-administering the prostaglandin receptor antagonist laropiprant. Taking niacin with meals also helps reduce this side effect. Acquired tolerance will also help reduce flushing; after several weeks of a consistent dose, most patients no longer experience flushing. Reduction of flushing focuses on altering or blocking the prostaglandin mediated pathway. Slow- or "sustained"-release forms of niacin have been developed to lessen these side effects. One study showed the incidence of flushing was significantly lower with a sustained-release formulation, though doses above 2 g per day have been associated with liver damage, in particular, with slow-release formulations.

Prostaglandin (PGD_2) is the primary cause of the flushing reaction, with serotonin appearing to have a secondary role in this reaction. The effect is mediated by prostaglandin E2 and D2 due to GPR109A activation of epidermal Langerhans cells and keratinocytes. Langerhans cells use cyclooxygenase type 1 (COX-1) for PGE2 production and are more responsible for acute flushing, while keratinocytes are COX-2 dependent and are in active continued vaso-dilation. Flushing was often thought to involve histamine, but histamine has been shown not to be involved in the reaction.

Gastrointestinal and Hepatic

Gastrointestinal complaints, such as indigestion, nausea and liver failure, have also been reported. Hepatotoxicity is possibly related to metabolism via amidation resulting in NAD production. The time-release form has a lower therapeutic index for lowering serum lipids relative to this form of toxicity.

Metabolic

Although high doses of niacin may elevate blood sugar, thereby worsening diabetes mellitus, recent studies show the actual effect on blood sugar to be only 5–10%. Patients with diabetes who continued to take anti-diabetes drugs containing niacin did not experience major blood glucose changes. Thus overall, niacin continues to be recommended as a drug for preventing cardiovascular disease in patients with diabetes.

Hyperuricemia is another side effect of taking high-dose niacin, and may exacerbate gout.

Other

Side effects of cardiac arrhythmias have also been reported. Increased PT and decreased platelet count have been reported, therefore these should be monitored closely in patients who are also taking anticoagulants.

Particularly the time-release variety, at extremely high doses, can cause acute toxic reactions. Extremely high doses of niacin can also cause niacin maculopathy, a thickening of the macula and retina, which leads to blurred vision and blindness. This maculopathy is reversible after niacin intake ceases.

In Pregnancy

Niacin in doses used to lower cholesterol levels has been associated with birth defects in laboratory animals, with possible consequences for infant development in pregnant women.

Pharmacology

Pharmacodynamics

Niacin's therapeutic effects are partly mediated through the activation of G protein-coupled receptors, including niacin receptor 1 (NIACR1) and niacin receptor 2 (NIACR2) which are highly expressed in adipose tissue, spleen, immune cells and keratinocytes but not in other expected organs such as liver, kidney, heart or intestine. NIACR1 and NIACR2 inhibit cyclic adenosine monophosphate (cAMP) production and thus fat breakdown in adipose tissue and free fatty acids available for liver to produce triglycerides and very-low-density lipoproteins (VLDL) and consequently low-density lipoprotein (LDL) or "bad" cholesterol. Decrease in free fatty acids also suppress hepatic expression of apolipoprotein C3 (APOC3) and PPARg coactivator-1b (PGC-1b) thus increase VLDL turn over and reduce its production. It also inhibits diacylglycerol acyltransferase-2 (important hepatic TG synthesis).

The mechanism behind increasing HDL is not totally understood but it seems to be done in various ways. Niacin increases apolipoprotein A1 levels due to anti catabolic effects resulting in higher reverse cholesterol transport. It also inhibits HDL hepatic uptake, down-regulating production of the cholesterol ester transfer protein (CETP) gene. Finally, it stimulates the ABCA1 transporter in monocytes and macrophages and up-regulates peroxisome proliferator-activated receptor γ results in reverse cholesterol transport.

It reduces secondary outcomes associated with atherosclerosis, such as low density lipoprotein cholesterol (LDL), very low-density lipoprotein cholesterol (VLDL-C), and triglycerides (TG), but increases high density lipoprotein cholesterol (HDL). Despite the importance of other cardiovascular risk factors, high HDL was associated with fewer cardiovascular events independent of LDL reduction. Other effects include anti-thrombotic and vascular inflammation, improving endothelial function, and plaque stability. Adipokines are the adipocytes' produced mediators. Some adipokines such as tumor necrosis factor (TNF)-a, interleukins and chemokines, have pro-inflammatory effect and some others such as adiponectin have anti-inflammatory effect that regulates inflammatory process, decrease vascular progression and atherosclerosis.

Niacin also appears to upregulate brain-derived neurotrophic factor (BDNF) and tropomyosin receptor kinase B (TrkB) expression.

Research has been able to show the function of niacin in the pathway lipid metabolism. It is seen that this vitamin can decrease the synthesis of apoB-containing lipoproteins such as VLDL, LDL, IDL and Lipoprotein (a) via several mechanisms: (1) Directly inhibiting the action of DGAT2, a key enzyme for triglyceride synthesis; (2) It has the ability to bind to the receptor HCAR2 thereby decreasing lipolysis and FFA flux to the liver for triglyceride synthesis; and (3) increased apoB catabolism. On the other hand, HDL cholesterol levels are increased by niacin through direct and indirect pathways. (4) Niacin decreases CETP mass and activity, and this synergistic effect with the decrease in triglyceride levels, can indirectly raise HDL cholesterol levels. The study has also been able to show direct effects on the beta chain of ATP synthase (5) and on production (6) and hepatic uptake (7) of apoA-I also increase HDL cholesterol levels. Thus by affecting the pathway reducing lipid levels help in reducing CVD.

Pharmacokinetics

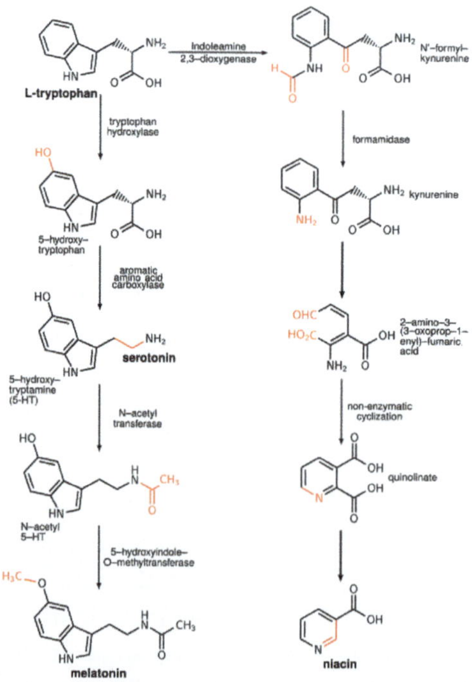

Niacin, serotonin (5-hydroxytryptamine), and melatonin biosynthesis from tryptophan

Biosynthesis

The liver can synthesize niacin from the essential amino acid tryptophan, requiring 60 mg of tryptophan to make one mg of niacin. Riboflavin, vitamin B_6 and iron are required in some of the reactions involved in the conversion of tryptophan to NAD.

Physical and Chemical Properties

Laboratory Synthesis

Several thousand tons of niacin are manufactured each year, starting from 3-methylpyridine.

Preparations

Many preparations of niacin are available over-the-counter as dietary supplements. Non-prescription extended release niacin, such as Endur-acin, which uses a wax matrix to delay release is available as well.

Over-the counter plain niacin has FDA regulatory approval. Some "no flush" types, such as inositol hexanicotinate contain convertible niacin compounds, but have little clinical efficacy in reducing cholesterol levels. "Slow release" varieties have higher hepatotoxic activity, hence some types of prescription niacin are not recommended due to potential harm.

Nicotinic Acid

Extended Release

A prescription extended release niacin, Niaspan, has a film coating that delays release of the niacin, resulting in an absorption over a period of 8–12 hours. The extended release formulations generally reduce vasodilation and flushing side effects, but increase the risk of hepatotoxicity compared to the immediate release forms.

A formulation of laropiprant (Merck & Co., Inc.) and niacin had previously been approved for use in Europe and marketed as Tredaptive. Laropiprant is a prostaglandin D2 binding drug shown to reduce vasodilatation and flushing up to 73%. The HPS2-THRIVE study, a study sponsored by Merck, showed no additional efficacy of Tredaptive in lowering cholesterol when used together with other statin drugs, but did show an increase in other side effects. The study resulted in the complete withdrawal of Tredaptive from the international market.

Inositol Hexanicotinate

One form of dietary supplement is inositol hexanicotinate (IHN), which is inositol that has been esterified with niacin on all six of inositol's alcohol groups. IHN is usually sold as "flush-free" or "no-flush" niacin in units of 250, 500, or 1000 mg/tablets or capsules. It is sold as an over-the-counter formulation, and often is marketed and labeled as niacin, thus misleading consumers into thinking they are getting the active form of the medication. While this form of niacin does not cause the flushing associated with the immediate-release products, the evidence that it has lipid-modifying functions is disputed. As the clinical trials date from the early 1960s (Dorner,

Welsh) or the late 1970s (Ziliotto, Kruse, Agusti), it is difficult to assess them by today's standards. One of the last of those studies affirmed the superiority of inositol and xantinol esters of nicotinic acid for reducing serum free fatty acid, but other studies conducted during the same period found no benefit. Studies explain that this is primarily because "flush-free" preparations do not contain any free nicotinic acid. A more recent placebo-controlled trial was small (n=11/group), but results after three months at 1500 mg/day showed no trend for improvements in total cholesterol, LDL-C, HDL-C or triglycerides. Thus, so far there is not enough evidence to recommend IHN to treat dyslipidemia.

Nicotinamide

Nicotinamide may be obtained from the diet where it is present primarily as NAD+ and NADP+. These are hydrolysed in the intestine and the resulting nicotinamide is absorbed either as such, or following its hydrolysis to nicotinic acid. Nicotinamide is present in nature in only small amounts. In unprepared foods, niacin is present mainly in the form of the cellular pyridine nucleotides NAD and NADP. Enzymatic hydrolysis of the co-enzymes can occur during the course of food preparation. Boiling releases most of the total niacin present in sweet corn as nicotinamide (up to 55 mg/kg).

Nicotinamide may be toxic to the liver at doses exceeding 3 g/day for adults.

Food Sources

Niacin is found in variety of foods, including liver, chicken, beef, fish, cereal, peanuts, and legumes, and is also synthesized from tryptophan, an essential amino acid found in most forms of protein.

Animal products:

- liver, heart and kidney (9 – 15 mg niacin per 100 grams)
- chicken, chicken breast (6.5 mg)
- beef (5 – 6 mg)
- fish: tuna, salmon, halibut (2.5 – 13 mg)
- eggs (0.1 mg)
- venison (8.43 mg)

Fruits and vegetables:

- avocados (1 mg niacin per 100 grams)
- dates (2 mg)
- tomatoes (0.7 mg)
- leaf vegetables (0.3 – 0.4 mg)
- broccoli (0.6 mg)
- carrots (0.3 – 0.6 mg)
- sweet potatoes (0.5 – 0.6 mg)

- asparagus (0.4 mg)

Seeds:

- nuts (2 mg niacin per 100 grams)
- whole grain products (4 – 29.5 mg)
- legumes (0.4 – 16 mg)
- saltbush seeds

Fungi:

- mushrooms, shiitake mushrooms (3.5 – 4 mg niacin per 100 grams)
- brewer's yeast (36 mg)

Other:

- beer (6 mg per pint, less if filtered)
- Ovaltine (18 mg)
- Peanut butter (15 mg)
- Tofu
- Soy sauce (0.4 mg)
- Vegemite (from spent brewer's yeast) (50 mg niacin per 100 grams)
- Marmite (from spent brewer's yeast) (50 mg niacin per 100 grams)

History

Niacin was first described by chemist Hugo Weidel in 1873 in his studies of nicotine. The original preparation remains useful: The oxidation of nicotine using nitric acid. For the first time, niacin was extracted by Casimir Funk, but he thought that it was thiamine and due to the discovered amine group he coined the term "vitamine". Niacin was extracted from livers by biochemist Conrad Elvehjem in 1937, who later identified the active ingredient, then referred to as the "pellagra-preventing factor" and the "anti-blacktongue factor." Soon after, in studies conducted in Alabama and Cincinnati, Dr. Tom Spies found that nicotinic acid cured the sufferers of pellagra.

Niacin is referred to as vitamin B_3 because it was the third of the B vitamins to be discovered. It has historically been referred to as "vitamin PP", "vitamin P-P" and "PP-factor", that are derived from the term "pellagra-preventive factor". When the biological significance of nicotinic acid was realized, it was thought appropriate to choose a name to dissociate it from nicotine, to avoid the perception that vitamins or niacin-rich food contains nicotine, or that cigarettes contain vitamins. The resulting name 'niacin' was derived from nicotinic acid + vitamin.

Carpenter found in 1951 that niacin in corn is biologically unavailable, and can be released only in very alkaline lime water of pH 11.

In 1955, Altschul and colleagues described niacin as having a lipid lowering property. Niacin is the oldest lipid lowering drug.

On April 22, 2014, Pennsylvania State University researchers working with NASA at the Goddard Astrobiology Analytical Laboratory reported of finding niacin within eight CM-2 type carbonaceous chondrite meteorites. The meteorite's vitamin B3 levels ranged from 30–600 parts per billion, the study reports. Related molecules discovered in the meteorites include pyridine carboxylic acids and pyridine dicarboxylic acids.

Research

In animal models and in vitro, niacin produces marked anti-inflammatory effects in a variety of tissues – including the brain, gastrointestinal tract, skin, and vascular tissue – through the activation of NIACR1. Niacin has been shown to attenuate neuroinflammation and may have efficacy in treating neuroimmune disorders such as multiple sclerosis and Parkinson's disease. Unlike niacin, nicotinamide does not activate NIACR1, however both niacin and nicotinamide activate the G protein-coupled estrogen receptor (GPER) *in vitro*. GPER activation also appears to modulate inflammatory processes, but the clinical significance of this effect has not been established as of 2016.

Golden Rice

Golden rice (right) compared to white rice (left)

Golden rice is a variety of rice (*Oryza sativa*) produced through genetic engineering to biosynthesize beta-carotene, a precursor of vitamin A, in the edible parts of rice. It is intended to produce a fortified food to be grown and consumed in areas with a shortage of dietary vitamin A, a deficiency which is estimated to kill 670,000 children under the age of 5 each year.

Golden rice differs from its parental strain by the addition of three beta-carotene biosynthesis genes. The rice plant can naturally produce beta-carotene in its leaves, where it is involved in photosynthesis. However, the plant does not normally produce the pigment in the endosperm, where photosynthesis does not occur.

In 2005, Golden Rice 2 was announced, which produces up to 23 times more beta-carotene than the original golden rice. To receive the USDA's Recommended Dietary Allowance (RDA), it is estimated that 144 g of the high-yielding strain would have to be eaten. Bioavailability of the carotene from golden rice has been confirmed and found to be an effective source of vitamin A for humans.

Golden rice has met significant opposition from environmental and anti-globalization activists that claim that there are sustainable, long-lasting and more efficient ways to solve vitamin A deficiency that do not compromise food, nutrition and financial security. A study in the Philippines is aimed to evaluate the performance of golden rice, if it can be planted, grown and harvested like other rice varieties, and whether golden rice poses risk to human health. Data has not been released yet.

Golden Rice was one of seven winners of the 2015 Patents for Humanity Awards by the United States Patent and Trademark Office.

As of 2016, it is still in development.

History

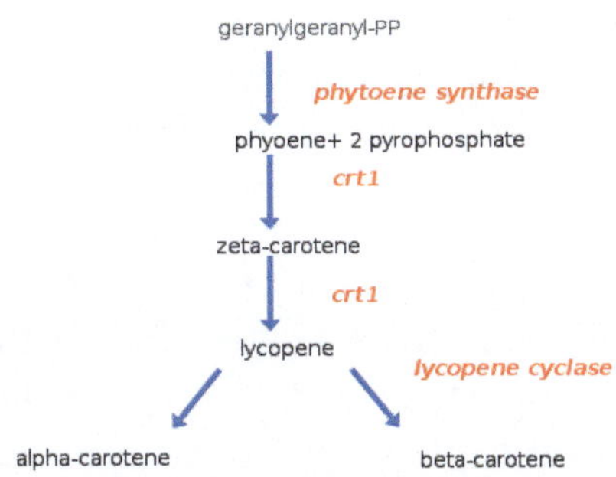

A simplified overview of the carotenoid biosynthesis pathway in golden rice. The enzymes expressed in the endosperm of golden rice, shown in red, catalyze the biosyntheis of beta-carotene from geranylgeranyl diphosphate. Beta-carotene is assumed to be converted to retinal and subsequently retinol (vitamin A) in the animal gut

Peter Bramley discovered in 1990s that a single phytoene desaturase gene (bacterial *CrtI*) can be used to produce lycopene from phytoene in GM tomato, rather than having to introduce multiple carotene desaturases that are normally used by higher plants. Lycopene is then cyclized to beta-carotene by the endogenous cyclase in Golden Rice.

The scientific details of the rice were first published in *Science* in 2000, the product of an eight-year project by Ingo Potrykus of the Swiss Federal Institute of Technology and Peter Beyer of the University of Freiburg. At the time of publication, golden rice was considered a significant breakthrough in biotechnology, as the researchers had engineered an entire biosynthetic pathway.

The first field trials of golden rice cultivars were conducted by Louisiana State University Agricultural Center in 2004. Additional trials have been conducted in the Philippines and Taiwan, and in Bangladesh (2015). Field testing provides an accurate measurement of nutritional value and enables feeding tests to be performed. Preliminary results from field tests have shown field-grown golden rice produces 4 to 5 times more beta-carotene than golden rice grown under greenhouse conditions.

Crossbreeding

In several countries, Golden rice has been bred with local rice cultivars. or crossbred with the American rice cultivar 'Cocodrie'. As of March 2016, golden rice has not yet been grown commercially, and backcrossing is still ongoing in current varieties to reduce yield drag.

Golden Rice 2

In 2005, a team of researchers at Syngenta produced Golden Rice 2. They combined the phytoene synthase gene from maize with *crt1* from the original golden rice. Golden rice 2 produces 23 times more carotenoids than golden rice (up to 37 µg/g), and preferentially accumulates beta-carotene (up to 31 µg/g of the 37 µg/g of carotenoids).

Genetics

Golden rice was created by transforming rice with two beta-carotene biosynthesis genes:

1. *psy* (phytoene synthase) from daffodil (*Narcissus pseudonarcissus*)
2. *crtI* (carotene desaturase) from the soil bacterium *Erwinia uredovora*

(The insertion of a *lcy* (lycopene cyclase) gene was thought to be needed, but further research showed it is already produced in wild-type rice endosperm.)

The *psy* and *crtI* genes were transferred into the rice nuclear genome and placed under the control of an endosperm-specific promoter, so that they are only expressed in the endosperm. The exogenous *lcy* gene has a transit peptide sequence attached, so it is targeted to the plastid, where geranylgeranyl diphosphate is formed. The bacterial *crtI* gene was an important inclusion to complete the pathway, since it can catalyze multiple steps in the synthesis of carotenoids up to lycopene, while these steps require more than one enzyme in plants. The end product of the engineered pathway is lycopene, but if the plant accumulated lycopene, the rice would be red. Recent analysis has shown the plant's endogenous enzymes process the lycopene to beta-carotene in the endosperm, giving the rice the distinctive yellow color for which it is named. The original golden rice was called SGR1, and under greenhouse conditions it produced 1.6 µg/g of carotenoids.

Vitamin A Deficiency

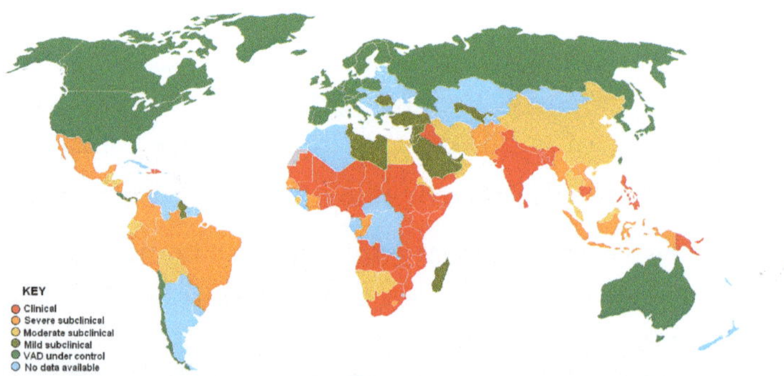

Prevalence of vitamin A deficiency. Red is most severe (clinical), green least severe. Countries not reporting data are coded blue. Data collected for a 1995 report.

The research that led to golden rice was conducted with the goal of helping children who suffer from vitamin A deficiency (VAD). In 2005, 190 million children and 19 million pregnant women, in 122 countries, were estimated to be affected by VAD. VAD is responsible for 1–2 million deaths, 500,000 cases of irreversible blindness and millions of cases of xerophthalmia annually. Children and pregnant women are at highest risk. Vitamin A is supplemented orally and by injection in areas where the diet is deficient in vitamin A.

As of 1999, 43 countries had vitamin A supplementation programs for children under 5; in 10 of these countries, two high dose supplements are available per year, which, according to UNICEF, could effectively eliminate VAD. However, UNICEF and a number of NGOs involved in supplementation note more frequent low-dose supplementation is preferable.

Because many children in VAD-affected countries rely on rice as a staple food, genetic modification to make rice produce the vitamin A precursor beta-carotene was seen as a simple and less expensive alternative to ongoing vitamin supplements or an increase in the consumption of green vegetables or animal products.

Initial analyses of the potential nutritional benefits of golden rice suggested consumption of golden rice would not eliminate the problems of vitamin A deficiency, but could complement other supplementation. Golden Rice 2 contains sufficient provitamin A to provide the entire dietary requirement via daily consumption of some 75g per day.

Since carotenes are hydrophobic, sufficient fat must be present in the diet for golden rice (or most other vitamin A supplements) to alleviate vitamin A deficiency. Vitamin A deficiency is usually coupled to an unbalanced diet. Moreover, this claim referred to an early cultivar of golden rice; one bowl of the latest version provides 60% of RDA for healthy children. The RDA levels advocated in developed countries are far in excess of the amounts needed to prevent blindness.

Research

Dr. José L. Domingo of the Laboratory of Toxicology and Environmental Health, School of Medicine, at Rovira i Virgili University in Spain said, "According to the information reported by the WHO, genetically modified products that are currently on the international market have all passed risk assessments conducted by national authorities." These assessments found no risk to human health. Domingo advocates continued research in the areas of GM rice and its effects on humans.

Clinical Trials/Food Safety and Nutrition Research

In 2009, results of a clinical trial of golden rice with adult volunteers from the US were published in the *American Journal of Clinical Nutrition.* The trial concluded that "beta carotene derived from golden rice is effectively converted to vitamin A in humans". A summary for the American Society for Nutrition suggested that "Golden Rice could probably supply 50% of the Recommended Dietary Allowance (RDA) of vitamin A from a very modest amount — perhaps a cup — of rice, if consumed daily. This amount is well within the consumption habits of most young children and their mothers".

It is well known that beta carotene is found and consumed in many nutritious foods eaten around the world, including fruits and vegetables. Beta carotene in food is a safe source of vitamin A.

The Food Allergy Resource and Research Program of the University of Nebraska undertook research in 2006 that showed the proteins from the new genes in golden rice showed no allergenic properties.

In August 2012, Tufts University and others published research on golden rice in the *American Journal of Clinical Nutrition* showing that the beta carotene produced by golden rice is as effective as beta carotene in oil at providing vitamin A to children. The study stated that "recruitment processes and protocol were approved". In 2015 the journal retracted the study, claiming that the researchers had acted unethically when providing Chinese children golden rice without their parents' consent.

Controversy

Critics of genetically engineered crops have raised various concerns. An early issue was that golden rice originally did not have sufficient vitamin A. This problem was solved by the development of new strains of rice. The speed at which vitamin A degrades once the rice is harvested, and how much remains after cooking are contested. However, a 2009 study concluded that golden rice is effectively converted into vitamin A in humans and a 2012 study that fed 68 children ages 6 to 8 concluded that golden rice was as good as vitamin A supplements and better than the natural beta-carotene in spinach.

Greenpeace opposes the use of any patented genetically modified organisms in agriculture and opposes the cultivation of golden rice, claiming it will open the door to more widespread use of GMOs. The International Rice Research Institute (IRRI) has emphasised the non-commercial nature of their project, stating that "None of the companies listed ... are involved in carrying out the research and development activities of IRRI or its partners in Golden Rice, and none of them will receive any royalty or payment from the marketing or selling of golden rice varieties developed by IRRI."

Vandana Shiva, an Indian anti-GMO activist, argued the problem was not the plant per se, but potential problems with poverty and loss of biodiversity. Shiva claimed these problems could be amplified by the corporate control of agriculture. By focusing on a narrow problem (vitamin A deficiency), Shiva argued, golden rice proponents were obscuring the limited availability of diverse and nutritionally adequate food. Other groups argued that a varied diet containing foods rich in beta carotene such as sweet potato, leaf vegetables and fruit would provide children with sufficient vitamin A. Keith West of Johns Hopkins Bloomberg School of Public Health has stated that foodstuffs containing vitamin A are often unavailable, only available at certain seasons, or too expensive for poor families in underdeveloped countries.

In 2008, WHO malnutrition expert Francesco Branca cited the lack of real-world studies and uncertainty about how many people will use golden rice, concluding "giving out supplements, fortifying existing foods with vitamin A, and teaching people to grow carrots or certain leafy vegetables are, for now, more promising ways to fight the problem".

In 2013, author Michael Pollan, who had critiqued the product in 2001, unimpressed by the benefits, expressed support for the continuation of the research.

Support

The Bill and Melinda Gates Foundation supports the use of genetically modified organisms in agricultural development. Specifically, the foundation is supporting the International Rice Research Institute in developing Golden Rice, a genetically modified rice variant used to combat Vitamin A deficiency.

In June 2016, 107 Nobel laureates signed a letter urging Greenpeace and its supporters to abandon their campaign against GMOs, and against Golden Rice in particular.

Protests

On August 8, 2013 an experimental plot of golden rice being developed at IRRI in the Philippines was uprooted by protesters. British author Mark Lynas reported in Slate that the vandalism was carried out by a group of activists led by the extreme left-inclined *Kilusang Magbubukid ng Pilipinas* (KMP) (unofficial translation: *Farmers' Movement of the Philippines*), to the dismay of other protesters. No local farmers participated in the uprooting; only the small number of activists damaged the golden rice crops because the farmers believe local customs which imply that killing a living rice plant is unlucky.

Distribution

Potrykus has enabled golden rice to be distributed free to subsistence farmers. Free licenses for developing countries were granted quickly due to the positive publicity that golden rice received, particularly in *Time* magazine in July 2000. Monsanto Company was one of the first companies to grant free licences.

The cutoff between humanitarian and commercial use was set at US$10,000. Therefore, as long as a farmer or subsequent user of golden rice genetics does not make more than $10,000 per year, no royalties need to be paid. In addition, farmers are permitted to keep and replant seed.

Dietary Supplement

A dietary supplement is intended to provide nutrients that may otherwise not be consumed in sufficient quantities.

Supplements as generally understood include vitamins, minerals, fiber, fatty acids, or amino acids, among other substances. U.S. authorities define dietary supplements as foods, while elsewhere they may be classified as drugs or other products.

There are more than 50,000 dietary supplements available. More than half of the U.S. adult population (53% - 55%) consume dietary supplements with most common ones being multivitamins.

These products are not intended to prevent or treat any disease and in some circumstances are dangerous, according to the U.S. National Institutes of Health. For those who fail to consume a balanced diet, the agency says that certain supplements "may have value."

Most supplements should be avoided, and usually people should not eat micronutrients except people with clearly shown deficiency. Those people should first consult a doctor. An exception is vitamin D, which is recommended in Nordic countries due to weak sunlight.

Definition

According to the United States Food and Drug Administration (FDA), dietary supplements are products which are not pharmaceutical drugs, food additives like spices or preservatives, or conventional food, and which also meet any of these criteria:

The product is intended to supplement a person's diet, despite it not being usable as a meal replacement.

The product is or contains a vitamin, dietary element, herb used for herbalism or botanical used as a medicinal plant, amino acid, any substance which contributes to other food eaten, or any concentrate, metabolite, ingredient, extract, or combination of these things.

The product is labeled as a dietary supplement.

In the United States, the FDA has different monitoring procedures for substances depending on whether they are presented as drugs, food additives, food, or dietary supplements. Dietary supplements are eaten or taken by mouth, and are regulated in United States law as a type of food rather than a type of drug. Like food and unlike drugs, no government approval is required to make or sell dietary supplements; the manufacturer checks the safety of dietary supplements but the government does not; and rather than requiring risk–benefit analysis to prove that the product can be sold like a drug, risk–benefit analysis is only used to petition that food or a dietary supplement is unsafe and should be removed from market.

Medical Uses

The intended use of dietary supplements is to ensure that a person gets enough essential nutrients.

Dietary supplements should not be used to treat any disease or as preventive healthcare. An exception to this recommendation is the appropriate use of vitamins.

Dietary supplements are unnecessary if one eats a balanced diet.

Supplements may create harm in several ways, including over-consumption, particularly of minerals and fat-soluble vitamins which can build up in the body. The products may also cause harm related to their rapid absorption in a short period of time, quality issues such as contamination, or by adverse interactions with other foods and medications.

Types

Dietary supplements being mixed into food for pigs.

There are many types of dietary supplements.

Vitamins

Vitamin is an organic compound required by an organism as a vital nutrient in limited amounts. An organic chemical compound (or related set of compounds) is called a vitamin when it cannot be synthesized in sufficient quantities by an organism, and must be obtained from the diet. Thus, the term is conditional both on the circumstances and on the particular organism. For example, ascorbic acid (vitamin C) is a vitamin for humans, but not for most other animals. Supplementation is important for the treatment of certain health problems but there is little evidence of benefit when used by those who are otherwise healthy.

Dietary Element

Dietary elements, commonly called "dietary minerals" or "minerals", are the chemical elements required by living organisms, other than the four elements carbon, hydrogen, nitrogen, and oxygen present in common organic molecules. The term "dietary mineral" is archaic, as the substances it refers are chemical elements rather than actual minerals.

Herbal Medicine

Herbal medicine is the use of plants for medicinal purposes. Plants have been the basis for medical treatments through much of human history, and such traditional medicine is still widely practiced today. Modern medicine recognizes herbalism as a form of alternative medicine, as the practice of herbalism is not strictly based on evidence gathered using the scientific method. Modern medicine, does, however, make use of many plant-derived compounds as the basis for evidence-tested pharmaceutical drugs, and phytotherapy works to apply modern standards of effectiveness testing to herbs and medicines that are derived from natural sources. The scope of herbal medicine is sometimes extended to include fungal and bee products, as well as minerals, shells and certain animal parts.

Amino Acids and Proteins

Amino acids are biologically important organic compounds composed of amine ($-NH_2$) and carboxylic acid ($-COOH$) functional groups, along with a side-chain specific to each amino acid. The key elements of an amino acid are carbon, hydrogen, oxygen, and nitrogen, though other elements are found in the side-chains of certain amino acids.

Amino acids can be divided into three categories: essential amino acids, non-essential amino acids, and conditional amino acids. Essential amino acids cannot be made by the body, and must be supplied by food. Non-essential amino acids are made by the body from essential amino acids or in the normal breakdown of proteins. Conditional amino acids are usually not essential, except in times of illness, stress, or for someone challenged with a lifelong medical condition.

Essential Fatty Acids

Essential fatty acids, or EFAs, are fatty acids that humans and other animals must ingest because the body requires them for good health but cannot synthesize them. The term "essential fatty acid" refers to fatty acids required for biological processes but does not include the fats that only act as fuel.

Bodybuilding Supplements

Bodybuilding supplements are dietary supplements commonly used by those involved in bodybuilding and athletics. Bodybuilding supplements may be used to replace meals, enhance weight gain, promote weight loss or improve athletic performance. Among the most widely used are vitamin supplements, protein drinks, branched-chain amino acids (BCAA), glutamine, essential fatty acids, meal replacement products, creatine, weight loss products and testosterone boosters. Supplements are sold either as single ingredient preparations or in the form of "stacks" – proprietary blends of various supplements marketed as offering synergistic advantages. While many bodybuilding supplements are also consumed by the general public their salience and frequency of use may differ when used specifically by bodybuilders.

Industry

In 2013, the global market of vitamins, minerals, and nutritional and herbal supplements (VMHS) was valued at $82 billion, with roughly 28 percent of that in the U.S., where sales increased by approximately $6 billion between 2007 and 2012.

The vitamins and dietary supplements sector in the U.S. grew 4% in 2015, to reach US$27.2 billion. The U.S. market was highly competitive in 2015, as no single company accounted for more than a 5% share of value sales.

Controversy

According to University of Helsinki food safety professor Marina Heinonen, more than 90% of dietary supplement health claims are incorrect. In addition, ingredients listed have been found to be different from the contents. For example, Consumer Reports reported unsafe levels of arsenic, cadmium, lead and mercury in several of the protein powders that were tested. Also, the CBC found that protein spiking (the addition of amino acid filler to manipulate analysis) was not uncommon, however many of the companies involved challenged their claim.

Adverse Effects

Nutraceutical

Some products make extraordinary claims and contain active ingredients which may not have been proven effective.

The number of incidents of liver damage from dietary supplements has tripled in a decade. Most of the products causing that effect were bodybuilding supplements. Some of the victims required liver transplants and some died. A third of the supplements involved contained unlisted steroids.

Mild to severe toxicity has occurred on many occasions due to dietary supplements, even when the active ingredients were essential nutrients such as vitamins, minerals or amino acids. This has been a result of adulteration of the product, excessive usage on the part of the consumer, or use by persons at risk for the development of adverse effects. In addition, a number of supplements contain psychoactive drugs, whether of natural or synthetic origin.

Physical and Chemical Properties

Adulteration in North America

BMC Medicine published a study on herbal supplements in 2013. Most of the supplements studied were of low quality, one third did not contain the active ingredient(s) claimed, and one third contained unlisted substances.

An investigation by the New York Attorney General's office reported in 2015 analyzed 78 bottles of herbal supplements from Walmart, Target, Walgreens and GNC stores in New York State using DNA barcoding, a method used to detect labeling fraud in the seafood industry. Only about 20% contained the ingredient on the label.

Some supplements were contamined by rodent feces and urine.

Only 0.3% of the 55,000 U.S. market dietary supplements have been studied regarding their common side effects.

Society and Culture

Public health

Work done by scientists in the early 20th century on identifying individual nutrients in food and developing ways to manufacture them raised hopes that optimal health could be achieved and diseases prevented by adding them to food and providing people with dietary supplements; while there were successes in preventing vitamin deficiencies, and preventing conditions like neural tube defects by supplementation and food fortification with folic acid, no targeted supplementation or fortification strategies to prevent major diseases like cancer or cardiovascular diseases have proved successful.

For example, while increased consumption of fruits and vegetables are related to decreases in mortality, cardiovascular diseases and cancers, supplementation with key factors found in fruits and vegetable, like antioxidants, vitamins, or minerals, do not help and some have been found to be harmful in some cases. In general as of 2016, robust clinical data is lacking, that shows that any kind of dietary supplementation does more good than harm for people who are healthy and eating a reasonable diet but there is clear data showing that dietary pattern and lifestyle choices are associated with health outcomes.

As a result of the lack of good data for supplementation and the strong data for dietary pattern, public health recommendations for healthy eating urge people to eat a plant-based diet of whole foods, minimizing processed food, salt and sugar and to get exercise daily, and to abandon Western pattern diets and a sedentary lifestyle.

Legal Regulation

United States

The regulation of food and dietary supplements by the U.S. Food and Drug Administration is governed by various statutes enacted by the United States Congress and interpreted by the U.S. Food and Drug Administration ("FDA"). Pursuant to the Federal Food, Drug, and Cosmetic Act ("the Act") and accompanying legislation, the FDA has authority to oversee the quality of substances sold as food in the United States, and to monitor claims made in the labeling about both the composition and the health benefits of foods.

Substances which the FDA regulates as food are subdivided into various categories, including foods, food additives, added substances (man-made substances which are not intentionally introduced into food, but nevertheless end up in it), and dietary supplements. The specific standards which the FDA exercises differ from one category to the next. Furthermore, the FDA has been granted a variety of means by which it can address violations of the standards for a given category of substances.

Regulation in European Union

The European Union's Food Supplements Directive of 2002 requires that supplements be demonstrated to be safe, both in dosages and in purity. Only those supplements that have been proven to be safe may be sold in the bloc without prescription. As a category of food, food supplements cannot be labeled with drug claims but can bear health claims and nutrition claims.

The dietary supplements industry in the United Kingdom (UK), one of the 28 countries in the bloc, strongly opposed the Directive. In addition, a large number of consumers throughout Europe, including over one million in the UK, and various doctors and scientists, had signed petitions by 2005 against what are viewed by the petitioners as unjustified restrictions of consumer choice.

In 2004, along with two British trade associations, the Alliance for Natural Health (ANH) had a legal challenge to the Food Supplements Directive referred to the European Court of Justice by the High Court in London.

Although the European Court of Justice's Advocate General subsequently said that the bloc's plan to tighten rules on the sale of vitamins and food supplements should be scrapped, he was eventually overruled by the European Court, which decided that the measures in question were necessary and appropriate for the purpose of protecting public health. ANH, however, interpreted the ban as applying only to synthetically produced supplements—and not to vitamins and minerals normally found in or consumed as part of the diet.

Nevertheless, the European judges acknowledged the Advocate General's concerns, stating that there must be clear procedures to allow substances to be added to the permitted list based on scientific evidence. They also said that any refusal to add the product to the list must be open to challenge in the courts.

Research

Effects of most dietary supplements have not been determined in randomized clinical trials and manufacturing is lightly regulated; randomized clinical trials of certain vitamins and antioxidants have found increased mortality rates.

Nutraceutical

Chia is grown commercially for its seeds rich in α-linolenic acid.

Nutraceutical, a portmanteau of the words "nutrition" and "pharmaceutical", was coined in 1989 by Stephen L. DeFelice, founder and chairman of the Foundation of Innovation Medicine. The term is applied to products that range from isolated nutrients, dietary supplements and herbal products, specific diets and processed foods such as cereals, soups, and beverages.

Regulation

Nutraceuticals are treated differently in different jurisdictions.

Canada

Under Canadian law, a nutraceutical can either be marketed as a food or as a drug; the terms "nutraceutical" and "functional food" have no legal distinction, referring to "a product isolated or purified from foods that is generally sold in medicinal forms not usually associated with food [and] is demonstrated to have a physiological benefit or provide protection against chronic disease."

United States

The term "nutraceutical" has no meaning in US law. Depending on its ingredients and the claims with which it is marketed, a product is regulated as a drug, dietary supplement, food ingredient, or food.

International Sources

In the global market, there are significant product quality issues. Nutraceuticals from the international market may claim to use organic or exotic ingredients, yet the lack of regulation may compromise the safety and effectiveness of products. Companies looking to create a wide profit margin may create unregulated products overseas with low-quality or ineffective ingredients.

Market

A market research report produced in 2012 projected that the worldwide nutraceuticals market would reach US$250 billion by 2018, defining that market as "Dietary Supplements (Vitamins, Minerals, Herbals, Non-Herbals, & Others), and Functional Foods & Beverages"

Classification of Nutraceuticals

Nutraceuticals are products derived from food sources that are purported to provide extra health benefits, in addition to the basic nutritional value found in foods. Depending on the jurisdiction, products may claim to prevent chronic diseases, improve health, delay the aging process, increase life expectancy, or support the structure or function of the body.

Dietary Supplements

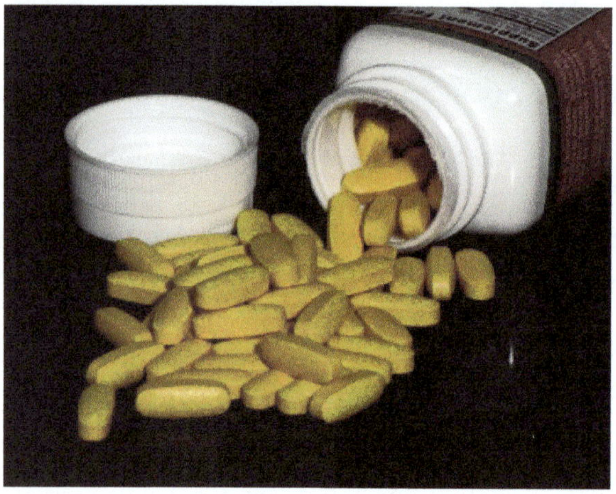

Dietary supplements, such as the vitamin B supplement shown above, are typically sold in pill form.

A dietary supplement is a product that contains nutrients derived from food products that are concentrated in liquid or capsule form. In the US, the Dietary Supplement Health and Education Act (DSHEA) of 1994 defined the term: "A dietary supplement is a product taken by mouth that contains a "dietary ingredient" intended to supplement the diet. The "dietary ingredients" in these products may include: vitamins, minerals, herbs or other botanicals, amino acids, and substances such as enzymes, organ tissues, glandulars, and metabolites. Dietary supplements can also be extracts or concentrates, and may be found in many forms such as tablets, capsules, softgels, gelcaps, liquids, or powders."

Dietary supplements do not have to be approved by the U.S. Food and Drug Administration (FDA) before marketing, but companies must register their manufacturing facilities with the FDA. With a few well-defined exceptions, dietary supplements may only be marketed to support the structure or function of the body, and may not claim to treat a disease or condition, and must include a label that says: "These statements have not been evaluated by the Food and Drug Administration. This product is not intended to diagnose, treat, cure, or prevent any disease."

Functional Foods

Functional foods are designed to allow consumers to eat enriched foods close to their natural state, rather than by taking dietary supplements manufactured in liquid or capsule form. Functional foods have been either enriched or fortified, a process called nutrification. This practice restores the nutrient content in a food back to similar levels from before the food was processed. Sometimes, additional complementary nutrients are added, such as vitamin D to milk.

Health Canada defines functional foods as “ordinary food that has components or ingredients added to give it a specific medical or physiological benefit, other than a purely nutritional effect.” In Japan, all functional foods must meet three established requirements: foods should be (1) present in their naturally-occurring form, rather than a capsule, tablet, or powder; (2) consumed in the diet as often as daily; and (3) should regulate a biological process in hopes of preventing or controlling disease.

History

Considered a father of Western medicine, Hippocrates advocated the healing effects of food.

The Indians, Egyptians, Chinese, and Sumerians are just a few civilizations that have used food as medicine. “Let food be thy medicine.” is a common misquotation attributed to Hippocrates, who is considered by some to be the father of Western medicine

The modern nutraceutical market began to develop in Japan during the 1980s. In contrast to the natural herbs and spices used as folk medicine for centuries throughout Asia, the nutraceutical industry has grown alongside the expansion and exploration of modern technology.

References

- Smith C, Lieberman M, Marks DB, Marks AD (2007). Marks' essential medical biochemistry. Hagerstwon, MD: Lippincott Williams & Wilkins. ISBN 0-7817-9340-8.
- Katzung, Bertram G. (2006). Basic and clinical pharmacology. New York: McGraw-Hill Medical Publishing Division. ISBN 0-07-145153-6.

- Brunton, Laurence L.; Lazo, John S.; Parker, Keith, eds. (2005). Goodman & Gilman's The Pharmacological Basis of Therapeutics (11th ed.). New York: McGraw-Hill. ISBN 0-07-142280-3.
- Staff (2009) Global Prevalence Of Vitamin A Deficiency in Populations At Risk 1995–2005 WHO Global Database on Vitamin A Deficiency. Geneva, World Health Organization, ISBN 978-92-4-159801-9, Retrieved 10 October 2011
- Robert S. Goodhart; Maurice E. Shils (1980). Modern Nutrition in Health and Disease (6th ed.). Philadelphia: Lea and Febinger. pp. 134–138. ISBN 0-8121-0645-8.
- Fitzgerald M (2014). Diet Cults: The Surprising Fallacy at the Core of Nutrition Fads and a Guide to Healthy Eating for the Rest of US. Pegasus Books. ISBN 978-1-60598-560-2.
- Hasler, Clare M. (2005). Regulation of Functional Foods and Nutraceuticals: A Global Perspective. IFT Press and Blackwell Publishing. ISBN 0-8138-1177-5.
- "Pellagra And Its Prevention And Control In Major Emergencies" (PDF). World Health Organization. World Health Organization. Retrieved 17 April 2015.
- Ahmad, Reaz (8 October 2015). "Bangladeshi scientists ready for trial of world's first 'Golden Rice'". The Daily Star. Retrieved 15 December 2015.
- Anahad O'Connor (February 3, 2015). "New York Attorney General Targets Supplements at Major Retailers" (the "Well" blog). The New York Times. Retrieved February 3, 2015.
- "What's wrong with herbal remedies They might not contain what they claim—and probably don't work, either". Consumer Reports. February 5, 2015. Retrieved February 7, 2015.
- "Nutraceuticals / Functional Foods and Health Claims on Foods: Policy Paper". Health Canada. June 24, 2013. Retrieved January 30, 2014.
- "Dietary supplement fact sheet: Folate". Health Information. Office of Dietary Supplements, US National Institutes of Health. Retrieved 11 January 2014.

Permissions

All chapters in this book are published with permission under the Creative Commons Attribution Share Alike License or equivalent. Every chapter published in this book has been scrutinized by our experts. Their significance has been extensively debated. The topics covered herein carry significant information for a comprehensive understanding. They may even be implemented as practical applications or may be referred to as a beginning point for further studies.

We would like to thank the editorial team for lending their expertise to make the book truly unique. They have played a crucial role in the development of this book. Without their invaluable contributions this book wouldn't have been possible. They have made vital efforts to compile up to date information on the varied aspects of this subject to make this book a valuable addition to the collection of many professionals and students.

This book was conceptualized with the vision of imparting up-to-date and integrated information in this field. To ensure the same, a matchless editorial board was set up. Every individual on the board went through rigorous rounds of assessment to prove their worth. After which they invested a large part of their time researching and compiling the most relevant data for our readers.

The editorial board has been involved in producing this book since its inception. They have spent rigorous hours researching and exploring the diverse topics which have resulted in the successful publishing of this book. They have passed on their knowledge of decades through this book. To expedite this challenging task, the publisher supported the team at every step. A small team of assistant editors was also appointed to further simplify the editing procedure and attain best results for the readers.

Apart from the editorial board, the designing team has also invested a significant amount of their time in understanding the subject and creating the most relevant covers. They scrutinized every image to scout for the most suitable representation of the subject and create an appropriate cover for the book.

The publishing team has been an ardent support to the editorial, designing and production team. Their endless efforts to recruit the best for this project, has resulted in the accomplishment of this book. They are a veteran in the field of academics and their pool of knowledge is as vast as their experience in printing. Their expertise and guidance has proved useful at every step. Their uncompromising quality standards have made this book an exceptional effort. Their encouragement from time to time has been an inspiration for everyone.

The publisher and the editorial board hope that this book will prove to be a valuable piece of knowledge for students, practitioners and scholars across the globe.

Index